J. W. (Hans) Niemantsverdriet, Jan Karel Felderhof
**Towards Scientific Leadership**

# Also of interest

*Compassionate Leadership.*
*For Individual and Organisational Change*
Kirstie Drummond Papworth, 2024
ISBN 978-3-11-076301-0, e-ISBN (PDF) 978-3-11-076312-6

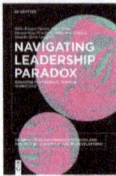

*Navigating Leadership Paradox.*
*Engaging Paradoxical Thinking in Practice*
Rikke Kristine Nielsen, Frans Bévort, Thomas Duus Henriksen, Anne-Mette
Hjalager and Danielle Bjerre Lyndgaard, 2023
ISBN 978-3-11-078885-3; e-ISBN (PDF) 978-3-11-078887-7

*De Gruyter Handbook of Organizational Conflict Management*
Edited by: LaVena Wilkin and Yashwant Pathak, 2022
ISBN 978-3-11-074601-3; e-ISBN (PDF) 978-3-11-074636-5

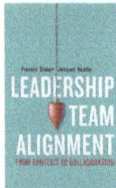

*Leadership Team Alignment.*
*From Conflict to Collaboration*
Frédéric Godart and Jacques Neatby, 2023
e-ISBN (PDF) 978-1-5036-3659-0

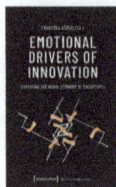

*Emotional Drivers of Innovation.*
*Exploring the Moral Economy of Prototypes*
Franziska Sörgel, 2024
e-ISBN (PDF) 978-3-8394-7147-0

J. W. (Hans) Niemantsverdriet,
Jan Karel Felderhof

# Towards Scientific Leadership

Personal Development Advice for Young Academics

2nd, Revised and Extended Edition

**DE GRUYTER**

**Authors**
Prof. Dr. J. W. (Hans) Niemantsverdriet
Syngaschem BV
Valeriaanlaan 16
5672 XD Nuenen
The Netherlands
jwn@syngaschem.com

Ir. Jan Karel Felderhof
Amersfoortsestraatweg 90 D003
1411 HG Naarden
The Netherlands
j.k.felderhof@gmail.com

ISBN 978-3-11-132531-6
e-ISBN (PDF) 978-3-11-132564-4
e-ISBN (EPUB) 978-3-11-132587-3

**Library of Congress Control Number: 2024943600**

**Bibliographic information published by the Deutsche Nationalbibliothek**
The Deutsche Nationalbibliothek lists this publication in the Deutsche Nationalbibliografie; detailed
bibliographic data are available on the Internet at http://dnb.dnb.de.

© 2024 Walter de Gruyter GmbH, Berlin/Boston
Cover image: natrot/iStock/Getty Images Plus
Typesetting: Integra Software Services Pvt. Ltd.

www.degruyter.com

# Foreword

This book results from a mutually inspiring collaboration for more than 30 years. The authors met for the first time in the early nineties: Jan Karel already established as a leadership development coach with programs in the Eindhoven area, focusing on how to transform authenticity into performance and impact; Hans, an associate professor at the Eindhoven University of Technology (TU/e), had just built an academic research group that was beginning to see some successes. They never lost contact ever since, and when Hans assumed an executive role as Dean at the university in 2001, he invited Jan Karel as his coach to be better prepared for the new "Deanly Duties."

The cooperation intensified in 2013 when Dr Yong-Wang Li, Founding Manager of Synfuels China, invited Hans to help build up a new, internationally oriented laboratory for fundamental research, SynCat@Beijing, with a local branch in the Netherlands, Syngaschem BV and to be the director of both. Jan Karel joined as the Director of Strategy, Organization, and Human Resources of Syngaschem BV. Together, they developed a course program on scientific leadership for the newly appointed scientific staff of the SynCat@Beijing Laboratory.

The authors organized five workshops in 2013–2014, two at the Boordhuys in Nuenen, The Netherlands, and three at Synfuels China's headquarters in Beijing-Huairou. Several guest speakers made excellent contributions: Dr. Yong-Wang Li, Dr. Jens Rostrup-Nielsen, Prof. Michael Bowker, and Prof. Gilbert Froment. Later, between 2016 and 2024, short courses were presented based on the book's first edition at the universities of Cardiff, Tianjin, Cape Town, the Free State, Darmstadt, and Vienna, and for the South African c*change network, the Turkish Catalysis Society, and the DIFFER Institute at Eindhoven.

This book is an extended version of the program that the authors developed for SynCat Ac@demy (www.scientificleaders.com). It finds its roots in Jan Karel's expertise in (self)-leadership since the 1990s in his company and the programs on presenting science that Hans developed at TU/e from 1990 to 2010.

https://doi.org/10.1515/9783111325644-202

# Acknowledgments

The authors are grateful to several colleagues, mentors, and friends with whom they developed ideas and course programs, which have become essential ingredients in the philosophy behind this book.

Jan Karel Felderhof wishes to thank:

- Dr. Martin Bakker, director at Océ Research, Venlo, and I designed a program on integrating technology development, product development, and product engineering from the viewpoint of research and market relevance.
- Ir. Piet Spohr, in his role as director of the TNO Physical and Electronics Laboratory, we developed ideas on how to focus a research institute of 400 scientists and engineers on projects and markets based on own science and technology.
- Ir. Poul Bakker, chairman of Company Coaching, and I set up talent programs together, particularly for the Philips Research Institute "NatLab."
- Professor Anton Hemerijck, director of the Dutch Scientific Council for Government Policies (WRR), Dean at the Free University of Amsterdam, and Professor at the London School of Economics, for collaboration on formulating theories for building stronger organizations based on trust, relevance, and self-confidence.
- Prof. Dr. Piet Cijsouw, director of the TU/e Postgraduate School, where we designed professional development programs for alumni and others.

Hans Niemantsverdriet is grateful to:

- Diek Koningsberger (then at TU/e) was instrumental in the development of our jointly created course "Oral and Poster Presentations" for PhD and MSc students at TU/e around 1990. The course was later published as a brochure and distributed by the European Federation of Catalysis Societies (EFCATS) in print and online. It serves as the basis for Chapter 3.
- Roel Prins (ETH Zurich) developed excellent lectures on writing scientific papers. Around 2010, Roel and Hans joined forces and created a combined video course, with the highly appreciated support of Elsevier's Lily Khidr and TU/e's Maurice Megens. It is still available on the internet (www.catalysiscourse.com).
- On leadership and management theory, Hans is greatly indebted to Ad Bossers and Joost van den Brekel at TU/e, who encouraged him to follow as many of their course programs as possible. Hans also acknowledges two Executive Chairmen at TU/e, Henk de Wilt and Amandus Lundqvist, who, each in his characteristic style, represented superb examples of authentic and visionary leadership in action.

Together, the authors express their sincere gratitude to:

- Yong-Wang Li, Founding Manager of Synfuels China Technology Co., Ltd, warmly embraced the idea of a SynCat Ac@demy with scientific leadership courses and provided generous financial support for the project. We are pleased to have two

https://doi.org/10.1515/9783111325644-203

contributions from him in the book. We also gratefully acknowledge the support of Yong Yang and Jian Xu.

- Jens Rostrup-Nielsen, who sadly passed away in 2022, for sharing his views on leading scientific research in industry and academia and for actively contributing to the SynCat courses. We are grateful for his friendship and mentorship in leadership within research environments. We are honored to have his views in a guest column in this book.
- Graham Hutchings. He and Hans presented courses for the c*change network in South Africa long before the idea of a book existed. More recently, they also offered courses for the Cardiff Catalysis Institute at Cardiff University. Together with both authors, the three presented a course at the University of Cape Town in January 2017. We warmly thank Graham for his guest column in this book.
- Michael Bowker. Mike actively contributed to several courses, provided constructive feedback, and occasionally warned us when we took it too far ("beware of the glassy eyes in the audience . . .").
- Michael Claeys and Jannie Swarts for organizing short courses and enthusiastically sharing their insights at the Universities of Cape Town and the Free State in Bloemfontein, respectively, at the beginning of 2017. We are grateful to Michael for his guest column in the book.
- Abhaya Datye helped us prepare for SynCat@Beijing and contributed his views on leadership in a guest column.
- Anton Hemerijck, a long-time collaborator of Jan Karel, for his guest contribution on impactful research in the social sciences.
- Karin Sora of De Gruyter, Berlin, a long-time friend in publishing Hans' books, for her warm support in this project. We are grateful for the opportunity to prepare an updated and expanded edition.
- The ambitious staff of SynCat@Beijing, Synfuels China Technology Co., Ltd., and SynCat@DIFFER, Syngaschem BV, for their active participation, contributions, and constructive critical feedback on the course programs. We acknowledge the valuable assistance and support of Antonio Vaccaro and the most valuable interactions with Foteini Sapountzi, Hans Fredriksson, Richard Gubo, Kees-Jan Weststrate, Ren Su, Xi Liu, Xiong Zhou, Ling Zhang, Chenghua Zhang, Xiadong Wen, and Jian Xu. It is a great privilege for us to have had the opportunity to interact closely with so many talented young scientists – the scientific leaders of the future!

Hans Niemantsverdriet and Jan Karel Felderhof,
Nuenen and Naarden, The Netherlands, and Laveno, Italy, June 2024.

# Special foreword

When Hans Niemantsverdriet and Jan Karel Felderhof asked me to write a few words of introduction to this book on scientific leadership, I read the manuscript with great interest. It is a great analysis which would have been very useful in my own situation many years ago, when I was struggling to start my career in research. Had I read this book, I could have done my job of the past 20 years more efficiently.

Over the past years and before this book was written, the authors and I worked closely together to establish the laboratories SynCat@Beijing and Syngaschem – SynCat@DIFFER. Jan Karel and Hans have done excellent work in helping our young post-doctoral fellows start their work and bring it to a higher level.

Young scientific researchers are facing many challenges, not only regarding their own scientific qualities, but also in adapting to the present competitive scientific climate; the latter is almost more influential for their careers. The funding system is one of the most critical factors. Young researchers have little other choice than to hunt for funds, and if they are successful, the grant is normally only small, say less than 200,000 US dollars. The odds are that they get discouraged after repeated disappointments and leave science, implying they have wasted valuable time and we do not benefit from their potential. I have personally addressed this several times in lectures for executives and policy makers. If scientists are totally controlled by the funding system, science will become a dead-end street!

Proper training in management skills and leadership makes young scientists better prepared to face and cope with these and other problems. Not only do they learn how to compete for scarce money, but also how to train oneself to get along well with people and interact effectively with colleagues, students, and leaders, and, importantly also, how to optimize one's performance in front of an audience and create beneficial publicity.

In my entire career, much of my working time, and some precious private time, has been spent on getting funding, managing affairs, and the most difficult challenge, interacting with people. I would have loved to devote much more time to my interests in science.

I wish I had known Jan Karel and Hans much earlier!

Yong-Wang Li
Professor Chinese Academy of Sciences
Founding Manager Synfuels China

https://doi.org/10.1515/9783111325644-204

# Contents

# 1 Creating success in your own scientific world

## 1.1 The world of research and development is highly competitive

**Let's assume you are a young academic.** You were appointed as an assistant or associate professor at a well-established university a few years ago or as a junior researcher in a research institute and you are working hard to become an acknowledged scientist in your research field. You have a dream about an important scientific challenge that you hope to solve in the coming years. You are qualified for your job because you hold a PhD degree from a good university, you successfully finished a couple of postdoc positions at internationally renowned research institutions, you have published together with famous scientists, you have several articles in high-impact journals, and you regularly present talks and posters at essential conferences. Your peers are starting to know and respect you; perhaps you already succeeded in getting grants from your national science organization to fund a substantial part of your research. You have a few PhD students and a postdoc working for you, each on their own exciting project. Your teaching load may be substantial, but the students appreciate your enthusiastic lectures, and some are eager to work with you on laboratory projects. In other words, it seems that you are well on your way in your scientific career.

**Not everything is going well, however.** You work long hours every day, and the balance between private and work life differs from what you, your partner, or your family would like. Daily issues in your work seem to determine your agenda and dominate your thoughts. Little irritations sometimes affect the atmosphere of your research group. Too often, you need more time to talk to your students regularly. Your head of department sometimes criticizes you for missing deadlines for important reports or other requests for information. The dean recently asked you to clarify the focus of your research, seen the variety of subjects you are covering. Also, it is not entirely clear what your group will be doing in a few years. This is also caused by the diversity of funding sources you have to deal with and the lack of consistent, sustained funding for a particular project, maybe even the one you like most.

**Experiences of first-year assistant professors**

A recent survey among 111 newly appointed assistant professors at mostly research-intensive (85 out of 111) American universities on their experiences after the first year revealed that 93% have worries about the current funding environment, while 81% feel pressure to present preliminary data or publish quickly.

As the most challenging aspects of the first year, the young professors mentioned managing time (34%), getting the laboratory going (23%), managing students and postdocs (15%), and teaching (10%).

Mentoring students was seen as the most rewarding task of their professorship by 52% of the respondents; 33% mentioned getting results in the laboratory, and 9%, teaching. Experienced faculty was by far the most important source of help and advice (62%); friends from graduate school, previous mentors, or others accounted for 12–13% each.

https://doi.org/10.1515/9783111325644-001

About 60% of the first-year professors felt that their work-life balance was less good than before, although 23% found that it had improved.

Source: New Professors by the Numbers, *Chemical & Engineering News*, CEN.ACS.ORG; American Chemical Society, May 22, 2017

This situation is typical for young academics – and sometimes for the more senior professors as well. As academics, we learnt very well how to do our research, explain things to others, or plan our research projects. However, insights in organization, strategic planning, managing, or even how to be or become a leader are usually much less developed, as only a few academic organizations pay sufficient attention to it. Here are some of the key questions:

– How do you organize all your activities, and how do you maintain a proper balance between them?
– How do you judge if you are on the right track to becoming a tenured professor, an influential scientist, or a reputed educator in your field of science?
– Are you sufficiently clear about what you want to achieve with your team, and how do you ensure that your team's activities create synergy?
– How can you become a true scientific leader, so that you can distinguish yourself with research that will have impact?
– And do you actually know how success is defined for your activities? Excellent papers in top journals? Prizes and awards? Your own future research institute? Having delivered several successful graduates that are quickly finding their way in society?

### 1.1.1 The academic climate has changed considerably over the past decades

Historically, universities were the exclusive seats of scientific research, which was done in close connection with teaching. Students were mentored by a professor, under whose personal supervision they studied and did research, experimental or theoretical, in a close mentor-pupil relationship. von Humboldt's ideal of academic freedom at independent universities was generally accepted; the role of the government was limited to providing funds and setting the legal framework for science and education. Once appointed, the chair holder or professor had full freedom to determine the research subjects. Teaching mainly was arranged within the professor-student mentorship and on a primarily individual basis, which later changed to educational programs organized by the faculty, the joint academic staff of a university department. The implicit understanding was that academics are intrinsically motivated to strive for excellence, and the climate was to "leave science to the academics." Scientific production, such as publications or dissertations, was not assessed by external parties in any way. Academic quality was merely how good the professor was in the eyes of their peers. An important factor was that access to academic education was limited to students who (or whose parents) could afford it, and in practice, the student/staff ratio was much smaller than it is now. Regarding scientific quality, the academic landscape had plateaus, several valleys – some deep – and incidentally peaks, and insiders knew very well where to go for scientific excellence.

**von Humboldt's ideal of higher education**
Wilhelm von Humboldt (1767–1835), philosopher, minister, and diplomat in Prussia, was a firm believer in academic freedom for professors and their students and in the integration of education, arts, sciences, and research, to offer comprehensive studies along cultural knowledge in a single institution. Here, students would be allowed to choose their own course of study, enabling them to become autonomous individuals and world citizens by developing their own reasoning powers in an unbiased environment of academic freedom.

He founded the University of Berlin, which was later named after his two-year younger brother Alexander. von Humboldt's concept was widely followed, first in North-Western Europe, and later also in the United States and other parts of the world.

In the twentieth century, and notably after the Second World War, science and research gradually were no longer the exclusive domain of universities but also of governmental or private research institutes and not to forget the industry. Companies like IBM, EXXON-Mobil, Siemens, Philips, AT&T, General Electric, Shell, and many others were known for their large corporate laboratories, which sometimes were deeply engaged in research of fundamental nature as well, and, as real powerhouses of scientific knowledge and development, almost functioned as an academic research institution.

In the academic world, universities saw the number of students rise, leading to much higher and less favorable student/staff ratios than before. Massive classes with several hundred first-year students in a lecture were no exception anymore. Funding did not increase proportionally, however, and scarcity of funds naturally led to the involvement of politics in decisions on how to distribute budgets over the various institutions. National funding agencies acting on behalf of the government started to play an increasingly larger role in the distribution of funds based on proposals. Hence, universities had to attract an increasingly growing part of their income through applications, of which only a fraction could be granted. Acquisition of funds from external sources in competition became the norm. To coordinate all this, the management of universities was professionalized to levels common in the business world. Management theories were introduced in which the distribution of resources is based on quantitative parameters to measure performance and output (the so-called key performance indicators), which are perceived to reflect the quality of research and teaching.

### 1.1.1.1 Striving for excellence

Around the turn of the century, 2000, "excellence" became the keyword in the global academic world. Many countries established "excellence" criteria and started to rank their universities accordingly. And if countries chose not to make such distinctions, then international organizations, magazines, or even universities themselves made up such rankings, based on criteria such as impact of publications, awards received, size of international cooperation network, participation in public-private partnerships, etc. As American top universities generally dominate such rankings, these institutions were increasingly seen as ideal examples of the perfect academic institution, not only in Asia – where this was traditionally the case anyway in countries such as Singapore and China – but more and more also in the alma mater of the academic world, Europe.

In parallel, the public opinion (or perhaps more precise, the opinion of some politicians and influential captains of industry) became an important factor as well. Universities were criticized for being inefficient and overly expensive, allowing students to take much longer than the nominal time to complete their studies, and for doing research that was insufficiently useful for society. In particular, European countries felt that their ability to innovate was weak compared to, e.g., the United States or emerging economies in Asia. Universities encouraged their staff and students to become entrepreneurs and to start companies. Several universities, together with local government, founded innovation centers, science parks, and incubators to facilitate start-ups and interaction with all sorts of companies. The European Union (EU) created many programs in which international consortia of academic and industry, particularly small-medium enterprises, could submit proposals for joint projects. The EU also started the European Institute of Technology, initially thought as a European MIT (modeled after the Massachusetts Institute of Technology in Boston) but later realized in the form of complex and largely virtual networks of Knowledge and Innovation Centres across Europe.

This is not to say that all of these trends are necessarily negative. Several successes of innovations starting from universities can be claimed. Of course, insight in entrepreneurial activities widens the horizon of students and may well prepare them better for their careers. However, universities should not forget what they are for, namely, education and training in science and technology at the highest possible level. In this sense, a famous book from the early 2000s appeared (Lewis 2005), entitled *Excellence Without a Soul – How a Great University Forgot Education*; in the language of the internet, "university.edu" became "university.com."

Another trend in the twenty-first century is that national or regional governments started "excellence initiatives," where universities or even clusters of universities could compete for star status and be recognized as "even more excellent than the rest." In some countries like the Netherlands, a significant part of the direct funding for academia was taken away and added to competitive funds of the national science organization. Successful initiatives received substantial financial support, but all in all, the success rates of proposals for individual academics decreased to alarming levels of 10% and below, implying that scientists are forced to spend a considerable part of their time to write proposals that are very likely to be rejected due to scarcity of funds, or raise funding in different ways. As proposals need to be reviewed by scientific peers, the pressure on the scientific world to review and assess proposals and participate in juries also increased considerably. The imbalance between highly rated proposals and available funds meant that many high-quality initiatives did not receive the means to realize them.

### 1.1.1.2 Publications, awards, and other academic currency

Excellence needs to be demonstrated. Publications are probably the most important academic currency in this respect, and their number is growing exponentially. The growth rate of cited scientific publications has risen from less than 1% before the middle of the eighteenth century to 2–3% in the first half of the twentieth century and 8–9% today, implying a doubling of publications in less than every 10 years (Bornmann & Muetz 2015). This is partly caused not only by the advent of research in emerging countries but also because of the "publish or perish" climate felt by academics all over the world, and not to forget the fact that many universities nowadays request students to publish at least one or more papers in established journals before they can submit their PhD thesis. Of course, all these submissions have to be handled by editors, reviewed by peers, and added to the activities of the academics. Who has time to read all these articles? Do we have to?

Prizes and awards form a second indicator of successful scientific performance. No one will question that a Nobel Prize winner is an exceptionally good scientist, and there are undoubtedly other awards with very rigorous and objective selection criteria. However, over the last decades, we have also seen the advent of a flood of prizes, awards, or recognitions for which the selection procedures are often doubtful and dominated by political considerations. In addition, professional societies, with the aim to offer opportunities to their members to distinguish themselves, have installed many so-called awards, which in essence are no more than a plaquette, or a modest stipend for visiting a conference or spending a few weeks on sabbatical leave. Universities have installed prestigious prizes to honor famous persons (not necessarily always scientists) just to benefit from the publicity that such an event generates. It is not at all uncommon to find scientists who hire professional assistance to propose themselves or their students for recognitions or – sadly but true – use their personal influence to be invited for prestigious invited lectures at leading conferences.

### 1.1.1.3 How do "academic enterprises" define success?

Anno 2025, many universities have developed into enterprises largely led by managers rather than by intrinsically motivated academics, and all these institutions are in competition for the best students, postdocs, academic staff, and funding. According to Binswanger (2014), the world of scientific research is dominated now by what he calls "artificially staged competitions" on many levels:
- Universities in national and international rankings of many sorts all do their best to prove that they are "excellent" (for whatever that means).
- Departments among each other within their university for resources or outside in their respective disciplines for scientific status.

- Research groups in or outside their universities and institutes, on the basis of publications in high-impact journals, citation scores, personal awards, invited lectures, or in nationally organized peer reviews, where each group receives scores on a scale of 1–5, for example.
- Students in their departments, for "Best Student of the Year Contests" and alike.

This is the climate in which you will have to find your way, whether you are a young assistant professor working hard to get tenure (a so-called tenure tracker), the newly appointed head of department, or an established scientist who has just been appointed as the new director of a research institute. You will have to live with the pressure of the race for excellence, but it is important to once in a while reflect on the priorities that really matter and ask yourself the question what success means for you (see Tab. 1.1). Of course, favorable indicators will result if your research group is successful in terms of the true values in the right column of the table. However, we see too many young scientists who are entirely focused on the short-term successes of good performance indicators, assuming that these are the road toward success in the long run. Understandably, they copy what they have seen from supervisors with similar philosophies, but nevertheless questionable as a pathway to sustained success.

**Tab. 1.1:** How do you define success (for the institution, yourself, and your funding organization)?

| Key performance indicators (KPIs) | True values |
| --- | --- |
| Number of scientific publications? | Quality of your graduates? |
| Papers in *Science* or *Nature*? | Contribution to "knowledge"? |
| H-index? | Recognition for quality of teaching? |
| Large-sized research group? | Well-equipped laboratory supported by one or more experienced technicians? |
| Number of prizes/awards? | Recognition for the high quality of a coherent research program? |
| Number of graduates from your laboratory? | Success of your alumni's careers? |
| Amount of funding or number of grants acquired? | |
| National or international ranking? | |

## 1.2 Trends in governance, control, management, and leadership

Scientists have always had an intuitive resistance against being told what to do. It is in their nature to discover, to go for the adventure of the unknown. It is almost impossible to plan for discovery, and hence, the typical project management idiom of Gantt charts, deliverables, milestones, or the complicated collaborative schemes that are asked so often nowadays in funding programs do not appeal to the average scientist.

Nevertheless, they have to play the game to survive in the present academic climate, and many found ways to cope with such systems and still keep capacity for true "blue-sky-discovery-type" research activities. At the same time, academic institutions, research organizations, and funding bodies adapted increasingly the governance and control principles from the entrepreneurial business world, as we sketched earlier. These principles themselves are far from static either, and it is useful to sketch briefly how these have changed over the past century.

### 1.2.1 Historic developments in management, governance, and control

Managing a production factory is something else than directing a bank, coaching a sports team, or leading a university. Different situations require different approaches to governance and control, as the following examples illustrate:
-   Production calls for a task-oriented approach; think, for example, about the way Henry Ford used the division-of-labor principle in the assembly line for producing the T-Ford in the beginning of the previous century. The management concept is actually based on hierarchy in activities and tasks, according to the theory of Frederick Winslow Taylor (1911), an engineer also known for his invention of carbon steel tools.
-   Administrative processes need a procedure-oriented approach, with pure bureaucracy as the ultimate form. In the early twentieth century, the German sociologist Max Weber saw bureaucracy as the most efficient and objective way to organize processes and hierarchy, with the benefits of maintaining order, optimizing efficiency, and avoiding favoritism.
-   Sales and project management were considered to best require a result-oriented approach, which lead to the principle of management by objectives or management by results, which we owe to Drucker (1954).
-   Science, as well as the domain of professional services, requires a knowledge/ know-how-oriented approach, as in Maister's (1993) description of the management of professionals. Weggeman and Hoedemakers (2014) actually advocate that professionals flourish most when subjected to a minimum of management – you can imagine that he is one of our favorite leadership gurus.
-   In this book we describe a self-leadership governance approach to lead scientists, leave as much freedom for creativity and self-steering as by nature the discipline calls for, but nevertheless have the process and organization in hand (Felderhof 2007).

Many of these approaches were quite focused on their own domains of application, and they did not always fully result in the desired outcomes. A basic shortcoming of the early theories was that people and human interactions did not feature explicitly. This started to become evident in the middle of the previous century. We cite a famous article by Tannenbaum and Schmidt (1958), in which they described the continuum between fully autocratic, boss-centered leadership to the fully democratic management style in which task and responsibilities are delegated to the staff (see Fig. 1.1). In essence, they explicitly acknowledged the value of teamwork. A few years later, Blake and Mouton (1964) proposed their famous managerial grid, an early assessment tool for management style based on two parameters: task orientation (performance) and attention for good human relationships. People management was established. However, leadership entails more than management of tasks and people. We discuss this at length in the coming chapters of this book.

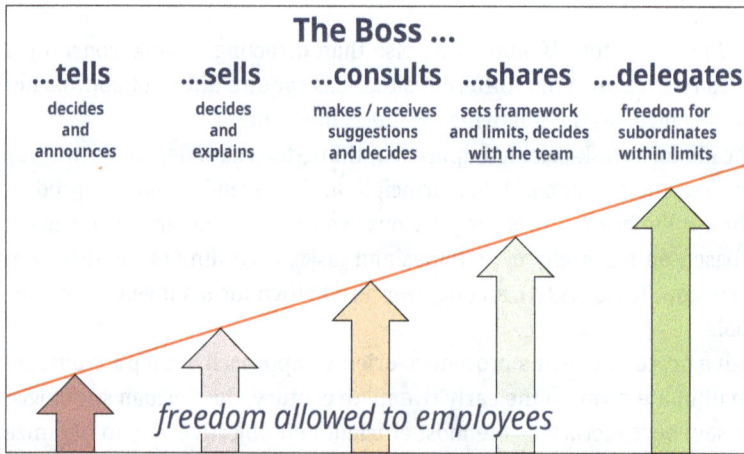

**The Boss ...**

| ...tells | ...sells | ...consults | ...shares | ...delegates |
|----------|----------|-------------|-----------|--------------|
| decides and announces | decides and explains | makes / receives suggestions and decides | sets framework and limits, decides with the team | freedom for subordinates within limits |

*freedom allowed to employees*

Fig. 1.1: How the boss deals with his personnel: Management style versus room for initiative and spontaneous action left for employees.

On the institutional level, governance and control theories initially concerned pure task management and control over results, while the state of the financial budget dominated the thinking. In the second generation, strategic planning according to the theories of Mintzberg (1978) and Porter (1980) played and increasingly important role. Governance and control interest focused monitoring the organization's progress in realizing its strategic plans and on the factors that determine financial outcome. The Balanced Scorecard was developed by Kaplan and Norton (1996) to keep track of success factors and key performance indicators. We described earlier how this thinking and methodology became common practice in the academic world. In essence, this second-generation governance and control is a top-down mechanism. It leaves little freedom for initiative and spontaneous creativity.

At present, we see in modern organizations, for example, in many start-ups, the transition to a third generation of governance and control (also referred to as "authentic generation"), which emphasizes the importance of comprehensive insight in the complexities that play a role (Felderhof 2007). Flexible (or agile) organizations require the intelligence and responsibility of people on several levels for making decisions on how to act in undefined, new situations. For this, people have to understand the goals and the needs of their organization thoroughly, and they need to dispose over sufficient freedom to operate, decide, and follow up. Terms like "self-steering," "self-directing, " "self-management," and even authentic "self-leadership" have entered the scene; hierarchical structures tend to become less dominant. In its ultimate form, third-generation governance and control rely on the unique abilities and strong points of individuals, who together operate like birds in a flock. The situation determines which abilities are needed most, and as in an effective flock of geese, the leadership changes accordingly in an almost automatic manner ("shared leadership" or "flock leadership").

Is it necessary to know all this if you want to become a successful researcher in a university, governmental, or industrial institute? We think having insight in the playing field and the rules of the game, whether these have been spelled out or are largely unwritten, and recognizing how institutions, be it your department, your university, the academy of sciences, or the company to which your institute belongs, are gradually changing their organization structures, will definitely help you in finding the way to achieve your goals.

### 1.2.2 The scientific leader of the early twenty-first century

We will define leadership later on in Chapter 2. For the moment, we assume that you are inspired by some brilliant idea, a greater plan, a dream, a vision of what you would like to achieve over time. Maybe it is solving a compelling problem, developing a new technology, sharing your enthusiasm for science with young students, or reaching the (perceived?) status of a successful scientist, that appeals to you. No matter if you only have it vaguely in your head or clearly spelled out in a few power point slides, there will be an inner driver behind where you are now.

Science is teamwork; very few scientists work alone. You will have students, postdocs, and perhaps assistants who work with you, and perhaps, you will collaborate with your colleagues. Most of your students were born after 2000. They belong to the so-called alpha generation, have probably been educated in a different world than you, and may have a different approach to learning (see Tab. 1.2), which presents some trends, differences, and expectations (Lemmens 2017, Lemmens et al. 2020).

The key to building a successful research group is to lead and inspire your students and coworkers, whatever their differences in approach and style, so that they get the most out of themselves while contributing optimally to your goals. That sounds obvious and is easily said, but how do you accomplish this?

**Tab. 1.2:** Characteristics of generations from the twentieth and twenty-first centuries and skills anticipated to become essential in the near future.

| Twentieth-century generations *"baby boomers, Gen X, Y, Z"* | Twenty-first-century generation *"generation alpha"* | Requirements for graduates of 2030 |
|---|---|---|
| *Verbal* | *Visual* | |
| Sit and listen | Try and see | Systems thinking |
| Teacher | Facilitator | Integrative and connecting |
| Job security | Flexibility | Transdisciplinary approach |
| Commanding | Collaborating | Creative |
| Curriculum-centered education | Learner-centered education | Entrepreneurial |
| Closed-book exams | Open-book world | Analysis *and* synthesis |
| Books and paper | Screens and devices | Global, mobile, agile |

> **My legacy in science are my graduates!**
> Asked what his legacy in science will be, 2017 Nobel Prize winner Prof. Fraser Stoddard answered: "My legacy will not necessarily be my chemistry, which has been described in more than 1,000 publications – too many! It will be the more than 400 graduate students and postdoctoral fellows whom I have trained and mentored. Almost 100 of them have gone on to be professors in their own right in universities all around the world, while many more have gone into industry, government, finance and publishing. It will be the young people I have trained who will be my legacy, particularly if they listen to my plea to tackle a big problem in science and not continue doing "Stoddard chemistry."
>
> From the interview by J.-F. Tremblay, *Chemical & Engineering News*, March 13, 2017, pp 34–35.

As we describe in detail hereafter, scientific leadership rests on three pillars: a compelling vision of what you want to achieve, relational skills to inspire your team and interact successfully with them, and managerial skills to organize everything that is needed to accomplish this.

In the early phase of your career, it is important to invest in the skills that help scientific leaders accomplish their job on a daily basis. Several of these are related to academic practices, which we will learn during our study time. Still, others also need your attention, for example, how to efficiently organize your work, manage your projects, teach your classes well, write articles and grant applications, and present with impact at conferences. Some chapters deal with these skills, and in particular, we stress the importance of presenting your work and ideas effectively in written and spoken form, by always having a clear message that will be well understood by the particular audience or readership that needs to understand the message and hopefully respond to it as you hope. These skills are commonly referred to as the "hard skills," or the "typical managerial skills." Mastering these is essential, but these skills alone do not guarantee successful leadership.

### 1.2.2.1 Academic managers or academic leaders?

It is important to appreciate the difference between management and leadership: Managers deal with tasks, keeping the business running, and keeping projects on track and often with a short- to medium-term focus. Leaders, on the other hand, focus on goals and the people who work hard to achieve these, with a long-term perspective. Hence, leadership includes all the management roles needed to keep the team on track, but it is definitely much broader than that. The comparison in Tab. 1.3 catches the difference.

Tab. 1.3: Differences between management and leadership.

| Managers | Leaders |
|---|---|
| Handle tasks | Motivate and inspire |
| Analyze data | Identify patterns and trends |
| Weigh alternatives | Handle the unpredictable |
| Solve problems | Judge how to cope with uncertainty |
| Take decisions | Support others to take decisions |
| Have often short-term focus | Have a long-term vision |

### 1.2.2.2 Relational skills, authenticity, and charisma

Some professors, group leaders, or directors seem to have been born as natural leaders; they give us the impression that they feel confident and at ease in everything they do. The question is if they feel it themselves too. Nevertheless, experience and insight acquired over the years, in a mix of successes and failures, contribute significantly to self-confidence and the ability to act authentically.

Authenticity and charisma are probably the most emphasized factors for successful leadership. These, however, are not characteristics that are easily acquired by reading books or following a miraculous course. These are characteristics that people possess to some extent and have learnt to develop further and rely on. You will probably feel best and be most successful in your work if you can behave according to your genuine personality, your "authentic self." It implies that you feel at ease, you know your strengths and weaknesses, and you communicate with others respectfully, in a relaxed and self-confident manner. Your self-confidence derives from knowledge and know-how, managerial and organizational skills, and insight into processes and human relations, which you achieved yourself or learnt from supervisors and mentors, combined with self-reflection and self-coaching.

Charisma is another often quoted attribute of "natural leaders." The term is not easy to define and is used in a variety of meanings, but let us, for simplicity, call it the ability to inspire others on the basis of warm personal relations. The word charisma comes from the Greek language and can be loosely translated as "giving in a graceful manner." Charismatic leaders have a natural talent for making others enthusiastic for their views and goals that often derive from a compelling vision. Can one learn how to become charismatic? We don't think so. You can learn to behave authentically, and if, as a leader, you believe in the power of synergy in your interaction with students, coworkers, and colleagues on the basis of a common ideal, where all can benefit in terms of their own goals as well, then chances are good that others may start to see you as charismatic. In other words, charisma will hopefully come and begin to be recognizable at some stage.

As a leader, you want to be an example for your students, coworkers, and even colleagues. Is there a secret recipe for becoming a scientific leader who inspires, educates, leads his/her team toward the realization of his dreams, and enables the team members to accomplish their goals? If there is a secret, it boils down to creating the environment and conditions under which your students, postdocs, and coworkers and yourself can flourish to achieve the best for themselves and simultaneously for you and your organization.

Our philosophy is that you can only be a true and inspiring leader if you have first learned how to lead yourself, i.e., feel confidently at ease in heading a group. As an accomplished and authentic self-leader, you will understand from experience how to inspire others and ensure that the people working with you progress through the same development toward authentic self-leadership. Is this a magic recipe? Not at all, we think it is mostly applying a healthy portion of common sense.

## References

Binswanger, M.: Excellence by Nonsense: The Competition for Publications in Modern Science. In: Opening Science; Bartling, S., Friesike, S. eds. Springer, Berlin, 2014; pp: 49–72. DOI: 10.1007/978-3-319-00026-8_3.

Blake, R., Mouton, J.: The Managerial Grid: The Key to Leadership Excellence; Gulf Publishing, Houston, 1964.

Bornmann, L., Muetz, R.: Growth rates of modern science: A bibliometric analysis based on the number of publications and cited references. J Assoc Inf Sci Technol; 2015; 66; 2215–2222.

Drucker, P.: The Practice of Management; Harper, New York, 1954.

Felderhof, J.P.K.: Enhancing Servant Leadership through Inspiring, Enabling and Focussing Enhancing Self-leadership; Informance Publishers, Eindhoven, 2007.

Kaplan, R.S., Norton, D.P.: The Balanced Score Card, Translating Strategy into Action; Harvard Business School Press, Boston, 1996.

Lemmens, A.M.C.: Eindhoven University of Technology, Private Communication, 2017.

Lemmens, A.M.C., van de Watering, G., Vinke, A.A., Rijk, K., Gómez Puente, S.M.: Engineers for the Future: Lessons Learned from the Implementation of a Curriculum Reform of TU/e Bachelor College. In: Proc. 48th Annual Conference on Engaging Engineering Education, SEFI 2020; Enschede, Netherlands, 2020.

Lewis, H.R.: Excellence without a Soul: How a Great University Forgot Education; PublicAffairs, Boston, 2005.

Maister, D.H.: Managing the Professional Service Firm; Simon & Schuster, New York, 1993.

Mintzberg, H.: The Structuring of Organisations; Pearson Education, London, 1978.

Porter, M.E.: Competitive Strategy: Techniques for Analyzing Industries and Competitors; The Free Press, New York, 1980.

Tannenbaum, A.S., Schmitt, W.H.: How to choose a leadership pattern. Harv Bus Rev; 1958; 36; 95–101.

Taylor, F.W.: Principles of Scientific Management; Harper, New York, 1911.

Weggeman, M.C.D.P., Hoedemakers, C.: Managing Professionals? Don't! How to Step Back to Go Forward: A Continental European Perspective; Warden Press, Amsterdam. 2014.

# Guest column: Define your passions and follow them to achieve long-term success

Abhaya Datye

*Dr. Datye is Distinguished Regents Professor of Chemical & Biological Engineering at the University of New Mexico, Albuquerque, NM, USA.*

As you start your career, you will be faced with many choices. Do you follow the path that leads to the highest salary in the near term, or should you focus on what is most satisfying? It is not an easy decision and those are not the only choices. Should you take up a job or start your own business? How about working for a nonprofit, or a start-up, or a large corporation? Each of these paths could lead to very different outcomes. While your educational journey has prepared you to embark on a career, as you reach the end of your graduate study, these difficult decisions start to appear. How best to make a decision that will lead you to be successful in the long term? Perhaps it is time to start thinking about what matters most to you, your passions. Because the path you follow is going to involve long-term commitment and hard work. So, it is even more important to think about what you are most passionate about. Identifying your passions will help you make the right decision at this stage.

It is possible that choices may be limited at the early stages of your career. You are going to choose the best path among those that are available. But do not despair if you must make a decision that appears suboptimal at the outset. There are learning opportunities in every path you follow, but you should remain focused on identifying your passions and making sure you stay committed. This is what will help you to stand out from a crowded field of initial hires in a larger corporation or in a smaller organization where you know everyone, but they may not know what special skills you bring to the team. As you aim for scientific leadership, it helps to define the focus of your work, the novel ideas you bring to the table, and your conception of the research landscape that helps you in identifying problems you will choose to study. Whether you are an independent investigator setting up an academic laboratory or you are part of a larger team, you have to think carefully about branding your work. As members of your team or department see your passion and commitment, they will start to recognize your achievements and give you the opportunity to move forward. Hence, it helps to be located at a position where your name comes up when people start to think who can be entrusted with leading a specific task or project.

## Toward a leading position

Let's think about the ingredients that will help you make strides toward leadership positions, first and foremost making connections and networking is very important.

This involves meeting researchers at conferences, especially after hearing their talks, or at social events. These interactions will give you a lot of insight into what is guiding the work in diverse organizations and to benchmark your work with those of others. Often, you will pick up more information from these chance encounters than you will gain from just reading the literature. It is here that you get to ask questions that probe deeper into the subject, and you will find that those you meet in these social settings may be much more conducive to sharing important information that they might never present at a podium or in published work. And, if you are in academia, some of the reviewers of your proposals or manuscripts will be right there in the audience. Hence, connecting with them will give you far better exposure than you might get by writing a manuscript. Tell them what you are passionate about and they will remember, and more opportunities will flow your way.

The second ingredient is to be well connected with literature. This requires using alerting services that send you the latest work published on a topic of interest to you. Make a point to catch up weekly with ongoing work in your field, which usually moves very fast. Staying abreast of the field also requires browsing journal TOCs and abstracts to learn about related developments that you may otherwise not learn about through a citation search. Teamwork is very important here, so getting your team members to also be doing the same sort of regular review of the literature is important, since it amplifies what you can do on your own. To achieve significant outcomes in research often takes a large team effort, so it is important to be engaged in not only your own professional development but also that of your team members. Ultimately, your success depends on each of your team members' contributions; they are just as important not only for doing the work, but for coming up with new ideas.

Finally, an important ingredient is to be active in service, both to your profession and also to your organization and to your community. This not only adds to the satisfaction you derive from your work but also gets you connected with people around you and elsewhere in the world. This is your chance to give up some of your time for the greater good, which is also an investment in the future. Professional service will get you into leadership positions and ultimately to recognition, which often arises as others get to see your passions and your dedication.

To summarize, achieving scientific leadership will require all the skills mentioned in this book. In addition, remember to stay true to your passions, since they will guide you toward a more satisfying work environment. Help your team members achieve success and give them credit. And most important, stay connected with literature and with the people who are doing related work worldwide. A scientific career is bound to lead to disappointments when something does not work. Use that as a learning opportunity. In the end, your scientific career will be much more rewarding if it is also satisfying the innate passions that shape your work.

# 2 Leading yourself and others in research

## 2.1 Introduction

In this chapter, we first analyze the sort of tasks that a researcher with a team of students and coworkers has to deal with, and we approach it from the viewpoint of the researcher self. We acknowledge that each person is different, while also the environments in which people work may vary widely. We adopt a simple four-dimensional model that has proven its value in describing the way different people act in their work, and we apologize at the onset for the severe oversimplification that we hereby introduce. Nevertheless, we believe the model provides valuable insight in how individuals may approach activities in their own manner. With two self-tests, you can score yourself on your personal performance dimensions, as well as on your approach to daily duties as expressed in your management dimensions. We end with a practical definition of leadership but stress that mastering self-leadership is essential for becoming a true leader. Without knowing oneself, one cannot understand and lead somebody else.

## 2.2 Your tasks as a research leader

Let us try to make an inventory of the various tasks that academics have to accomplish these days. Many of these will also be important for research leaders in institutes or industry, perhaps with the exception of teaching (although new team members need education, instruction, and encouragement to dive into the basics of their new activities as well). Often, universities assess their academic staff on three major aspects, namely, education, research, and organization, while sometimes contributions to the exploitation of research, innovation, and/or commercialization are also considered. We attempt to make a list (and anticipate that it may, on one hand, not be complete, while, on the other, not all these tasks may apply to you):

- Education: preparing and teaching classes, instructions, preparation of syllabi, handouts, exercises, question hours, preparing exam questions, correcting exercises, draft theses, grading exams, devising experimental practice courses along with written instructions, supervising, and mentoring students.
- Research: generating ideas and formulating research questions, reading literature, designing experiments, doing the actual research either by yourself or together with students, analyzing and interpreting results, writing reports and publications, going to conferences, preparing presentations, supervising research students and team members.
- Organization: obtaining funding for research (i.e., conceiving and writing convincing proposals), doing administrative duties, planning, reporting, arranging

https://doi.org/10.1515/9783111325644-002

group meetings and seminars, sitting on committees, networking, taking care of public relations.

– Exploitation (although perhaps limited to specific universities only): pursuing commercial exploitation of research, handling intellectual property issues (patents, etc.), devising a business plan for the initiation of a start-up.

All in all, this is an impressive list of activities, although not all may apply to you. Also note that many of the tasks in the list are typical management activities, in the categories that are typically based on "science" or more generally "content" and "knowledge," namely, teaching and research. Mastering skills as time management, project management, strategic planning, and relation management, together with "soft" skills to effectively supervise, coach and motivate people, and communicate with them through spoken and written word, is as important as the academically acquired scientific and educational skills. How do you cope with all these challenges and how do you ensure that each aspect or each task receives the attention it deserves? Everybody is different and has his/her own way in dealing with the complexities of life in academia or, more general, research and development. We think the following analysis helps in appreciating how different people deal with the complex tasks of managing their work.

## 2.3  Different ways in which people approach their work

We almost enter the field of psychology when we try to describe the ways and attitudes in which people approach their work and all the activities that belong to it. In Chapter 6, we go a bit deeper into the subject, but here, we will assume that people's behavior in work situations can be described by the four simple personal performance dimensions: feel, think, do, and drive (to be understood in the sense of achieve, perform, or deliver).

Figure 2.1 shows the four personal performance dimensions that we believe describe the way people manage their daily work: We all have the same blueprint in us for these dimensions and we developed them to variable extents. We describe them in the context of an academic research environment.

The *feeler* in us is empathic, sympathetic, and sensitive and reacts on the basis of intuition, inner wisdom, and first impressions. Feelings connect us with the subtleties of intangible intelligence. Emotions regard the positive (winning, success) or negative (loosing, frustration) reactions to activities or outbursts of suppressed feelings. Practical intuition gives us immediate directions on what to do in an imminent situation. Our feeling tells us whether such response is ethical and sustainable in the long term.

The feeler is strongly interested in personal interactions with others and is highly sensitive to emotions that first encounters with situations generate. He/she strongly appreciates the value of networking, good relations, and fruitful contacts with peers in the organization and the scientific or professional community. It is sometimes said

Fig. 2.1: The four personal performance dimensions, which we use as a simplified model for describing our behavior in the work place. Drive is meant to be understood as enforce, using the power of will to achieve.

that first impressions are determined by human emotions and feelings, before the ratio takes over.

The *thinker* in us likes all content-related aspects of the work, e.g., the intrinsic value of the research, the design or interpretation of new experiments, or what the literature has to say about a particular scientific question. This type of person may also like to think strategically about how meaningful activities are in a larger context or, for example, how the research and educational efforts can make a difference and will have impact. The thinker likes to carefully plan activities before they are done (see the cartoon in Fig. 2.2).

The *doer* thrives on being active, be it in the laboratory to do experiments or in the execution of tasks in general. If there is a plan that the doer believes in, he/she likes to carry it out. He/she usually values structure and organization, such as clear plans and protocols for experiments and standardized templates for reporting the results, as these provide clear and generally accepted frameworks for how things should be done. The doer is usually very active in accomplishing targets, networking, and professional activities such as education.

The *driver* (drive to be understood as the power of will to achieve) has focus on output and tangible results and gets energy from completing tasks and realizing goals and feels satisfied when a deadline has been met, or a report or paper has been submitted in time.

**Simplified picture of the four performance dimensions in a production company**
Allow us to sketch a stereotypical (and in fact too simple) division of roles in a company that designs and produces consumer goods. In the R&D department, the researchers are creative and strong in the *think* dimension. The developers are more focused on the technology and add the *drive* dimension (for in the end, it has to work). Engineers contribute the *do* dimension for the prototype that has to be tested and eventually produced. Designers supply the *feel* dimension, for clients will have to like the look, feel, and use of the product. Business developers and sales employees are again focused on the *drive* dimension to sell the product and make a profit.

**Fig. 2.2:** While the thinker is still thinking what needs to be done, the doer is already on the way to be active (reproduced with permission by Dan Reynolds).

## 2.3.1  Balance

Although each of us possesses these four characteristics *think–do–feel–drive* by nature in a certain mix, it is important that we are aware of our preferences and learn to compensate at least a bit for the ones that are too little present.

If we score ourselves on these four personal performance dimensions, we obtain graphs like the ones in Fig. 2.3. The balanced region is somewhere in the middle of each axis.

What do these graphs tell us? The person represented on the left of Fig. 2.3 thrives on doing things and has a strong drive to get things done. His/her scores on the feel and think dimensions are significantly lower. We could imagine that this profile corresponds to a manager who has a clearly described set of tasks, perhaps a production worker in a factory, or an analyst in a medical laboratory. The person on the right has a profile that looks more like a scientist, e.g., a professor with a large group of students. This person thinks a lot, clearly likes to interact with other people (the feel dimension), and has the inner drive to make sure that things will get done, but the students will have to do it mostly. We would call this a "balanced profile" because all dimensions are reasonably to strongly developed, without one being really too dominant.

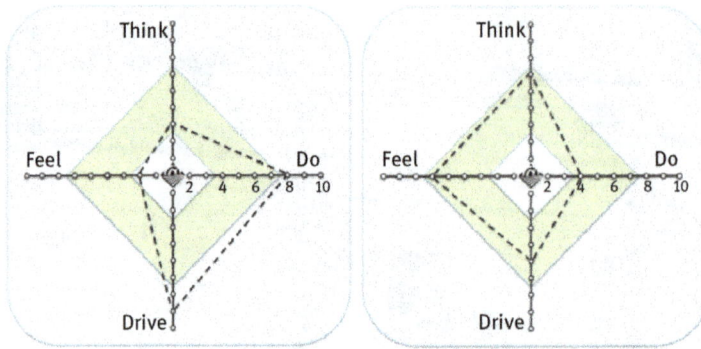

Fig. 2.3: Two examples of personal performance dimension scores. Left: A typical performer who will ensure that the organization produces predictable output and required results. Right: A more balanced profile, which could well be that of a successful professor who is fond of content and meaning and likes to interact with his/her people but at the same time feels a strong inner drive to realize his/her goals.

## 2.3.2 Knowing yourself

Knowing yourself and how you score on your personal performance dimensions is an important first step toward growth in your job. The Appendix has a scorecard for analyzing yourself and may be a starting point for self-reflection. Are your personal performance dimensions in accordance with the sort of position you have in mind? If you feel they are unbalanced, can you do something about it? You can also use the scheme to assess the people working with you. In a later chapter about human interactions, we will explain how this scheme may help you to recognize patterns in how people operate in teams.

There are, of course, many more important factors in understanding the way you work; we take Drucker's popular treatment on self-management as a guideline here (Drucker 2008). For example, recognizing the way in which you learn and deal with all kinds of information is an essential point to know. Do you mostly learn from reading texts, by listening, or perhaps by explaining things to others? Some people learn by taking extensive notes of every presentation they hear and write notebooks full during conferences and seminars because this is the way they absorb, process, and remember information (sometimes even without ever looking back to their notes). Our school systems are largely based on traditional ways in which a teacher explains new material and students process the new knowledge by listening, taking notes, studying books, and making exercises. However, many people learn much more effectively in other ways, for example, proactively, by doing, by trying out themselves, by searching the relevant information they need in the context of what they are busy with at that moment, and not because the professor or teacher tells it in a lecture that is delivered at a time that the student is perhaps not at his/her most perceptive state of mind anyways. So, it is vitally important that you know your own most effective ways to learn and those of your coworkers and students.

Other important questions about yourself are the following:

– Do you produce the best outcome by working alone, or together with others? You can, for example, still be a very successful leader if you prefer to do some essential tasks entirely by yourself, but it is important that you realize this and your team understands it, too.
– Are you good at taking decisions or by advising others on key matters? According to Drucker, the promotion of a successful second deputy to the director position often fails because an excellent advisor is not necessarily an effective decision maker. Having thought about this is obviously important before you decide to go for the highest leadership position.
– When do you perform at your best? Under conditions of challenge and stress, or do you need a more comfortable, predictable situation? Do you feel at ease in a large and complex organization, or rather in a smaller setting?
– It is an illusion to believe that you can change yourself easily on such fundamental aspects of your personality. Starting from your strong points and continuously working on enhancing your strengths and acquiring additional skills is a good strategy to build your career on. This will then be the basis for your (self) leadership.

People operating from a strong inner drive may be interested in taking a deeper look at themselves as the basis for growth. Louise Hay (1984) sold more than 30 million copies of her book on the deeper patterns underneath a personality and how you can change them. The authors have developed tests for giving insight in your self-leadership potential.[1]

## 2.4 People's achievement dimensions

The personal performance dimensions feel, think, do, and drive match well onto dimensions one encounters in the daily activities at work: interacting with people, content, structure, and performance, respectively (see Fig. 2.4).

The feeler is generally interested in other people and likes to interact with them; the thinker likes content and knowledge. Doers are helped when it is clear what needs to be done, as described in procedures and protocols, which all relate to clearly structured work activities. Self-propelling doers like being active with their profession, reaching targets, or networking. The driver likes to generate output and have impact, i.e., he/she is focused on achievements.

---

1 www.scientificleaders.com

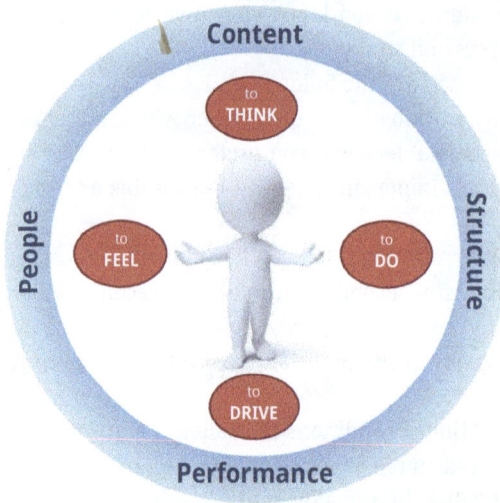

**Fig. 2.4:** Personal performance dimensions "feel, think, do, and drive" correspond to accomplishment dimensions "people, content, structure, and performance" (getting results). A well-functioning person is in control of realizing a successful performance in which results and overall output are in balance with content and relationships.

We can now add all the academic tasks to Fig. 2.4 and thus obtain Fig. 2.6. We immediately see a relationship between the type of activities various people will feel attracted to or may not be so good at. You can assess your own dimensions with the scorecards of Fig. 2.5, which are also available in the appendix.

## 2.5 Self-leadership

Now, what is an accomplished self-leader? It is somebody who has a good view on all what needs to be done, feels responsible for it, understands the relationship between character and personal inclinations for types of work, and has achieved a way to balance his/her personal performance dimensions with all the tasks that need to be accomplished. How? By learning the skills that he/she does not possess naturally and by tempering his/her natural inclinations to go for those aspects of the work that he/she feels most attracted to. It is a matter of learning the necessary skills, exploiting talents, relying on experience, and applying a mix of common sense, patience, managing your own emotions, and finding the right balance between all these factors. Above all, it is taking the responsibility for your group in your own hand and feeling the inner drive to act where it is needed. It is *your* research group, department, or institute, and this comes with an obligation; you cannot hide behind others.

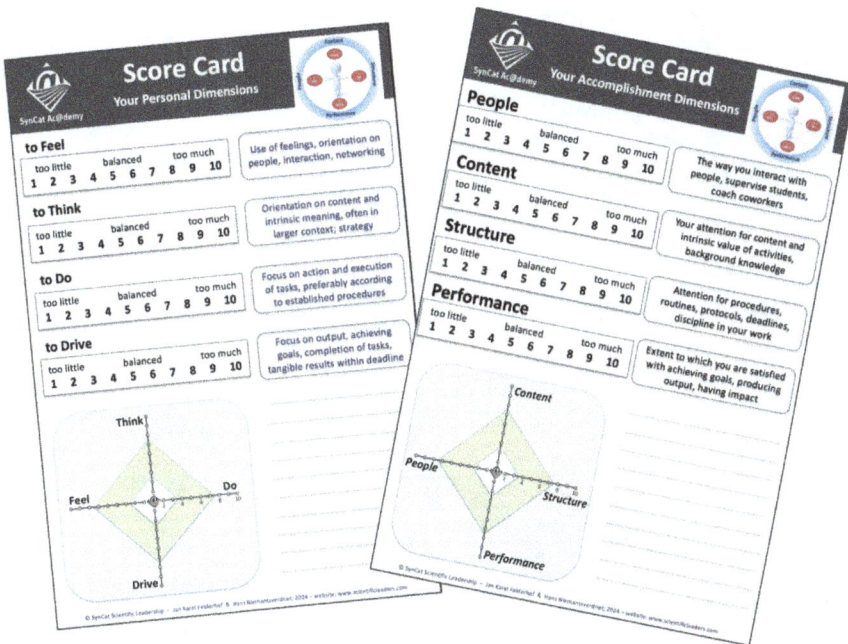

Fig. 2.5: The appendix has scorecards for personal and accomplishment dimensions.

**Self-leadership: history and definition**

"Self-leadership" as a notion was introduced in the scientific literature some 40 years ago by Manz (1983). He defined it a few years later as a "comprehensive self-influence perspective that concerns leading oneself toward performance of naturally motivating tasks as well as managing oneself to do work that must be done but is not naturally motivating" (Manz 1986). Another crisp and clear definition was given by Bryant and Kazan as "the learned ability to intentionally influence ourselves to achieve our objectives" (Bryant 2012). They stress that it forms the foundation for personal, team, business and strategic leadership.

Blanchard et al. (2005) formulated the responsibility part of self-leadership in a very compelling way:

*Organizations need people who act as if they own the place.*

These people dare to take initiative if they believe it is needed and feel personally responsible. They are empowered people who like to learn and develop further. They like to solve problems and they understand what is needed. And, very importantly, they are honest, sincere team players who have the confidence to be themselves and to behave authentically. This in contrast to people with a "victim mindset" who lack the courage to challenge what superiors say or to break through the constraints of "this is how things are done and have always been done in this organization." They

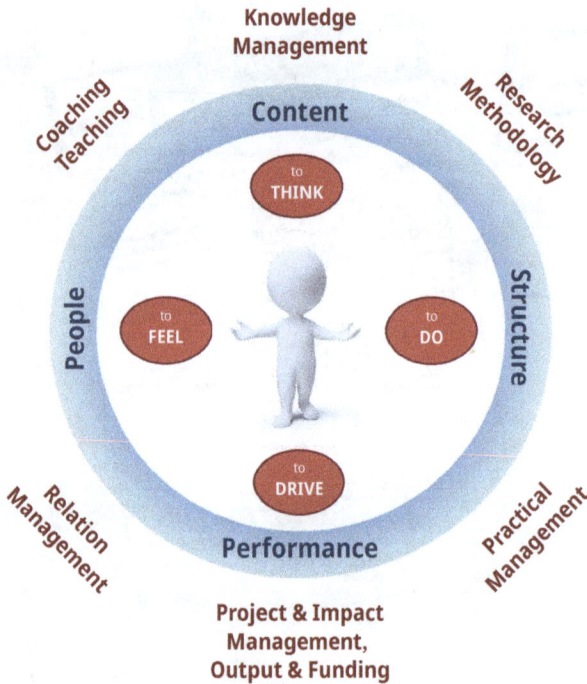

**Fig. 2.6:** Academics have to find a feasible balance between all the tasks involved in leading a research group. The person in the middle is you, the young academic, surrounded by all the activities you are engaged in. The four dimensions "think, do, drive, and feel" represent important aspects of your personality, which determine and energize the way you operate.

often assume that authority or seniority is needed to raise critical questions or to propose novel or alternative ideas.

The following quotation from Jack Welch, former chief executive officer (CEO) of General Electric, nicely illustrates the difference between victims and self-leaders:

> *You can look at the situation and feel victimized. Or you can look at it and be excited about conquering the challenges and opportunities it presents.*

Note also the subtle difference in how to involve someone's help: Do you ask "I have a problem, can you please help me" or do you say "I need someone who can . . ." The latter implies that you know what is needed, that you have a plan, while the former sounds rather helpless.

Does self-leadership derive from power or a higher position in the hierarchy? Not necessarily. It is in fact one of the most frequently assumed constraints to believe that one needs legitimized authority to ask others to assist or, in other words, to occupy a higher hierarchical position in the organization or have the specific mandate from a superior to involve others. Power in a generalized sense also derives from having specific knowledge, experience, skills, relationships, personality, previous successes, and so on.

Being an accomplished self-leader implies that within your span of control, you take maximal responsibility and initiative to ensure that things will go the way you think it is best for your team, your laboratory, your organization. Table 2.1, adapted from the work of Blanchard et al. (2005), summarizes the most salient characteristics of self-leadership.

Tab. 2.1: Summary of self-leadership (adapted from Blanchard et al.).

**Self-leaders take responsibility**

**Organizations need people who "act as if they own the place"**
- People that take initiative and feel responsible
- Empowered people who like to learn and develop
- Problem solvers who understand the needs
- Honest team-players who are their "authentic selves"

**Self-leaders challenge assumed constraints**

| | |
|---|---|
| "Assumed constraint": We have always done it this way . . . a belief based on past experience that limits you in your current situation. | Be constructively critical and open minded about ways to achieve goals. |

**Self-leaders celebrate their points of power**

| | |
|---|---|
| "Assumed constraint": Power derives from function in the organization: "I'm not in a position to get people to do what I want them to do . . ." | Recognize that there are various "sources of power": knowledge, relationships, personality, position, reputation, etc. |

**Self-leaders collaborate for success**

| | |
|---|---|
| Help me to solve my problem, please . . . | "I need someone who can . . ." implies a plan and shows initiative |

## 2.6  Toward scientific leadership – a practical definition

We return to Fig. 2.6. When you succeed in giving the required attention to a scientist's tasks in a way that works for you, it is time to consider how you are leading your group. If you are a relatively young assistant professor, your group may consist of a few graduate students. In the future, you may grow to a full professor with a large group of postdocs and students, and if you are so lucky to have them in your team, even some technicians and an administrative assistant. No matter if you are the novice or the experienced head of department, successful leadership implies that you succeed in creating an environment where people flourish and can perform at their best in such a way that they not only realize their own goals but also contribute to your goals or those of the organization that you are leading. Obviously, leadership belongs on the "people–feeling" side of the scheme, although it relates to every aspect of scientific life. Leadership is in essence how to realize goals with people, while you enable them to flourish and enhance themselves. It rests strongly on effective communication and organization as well. We have included both in Fig. 2.7.

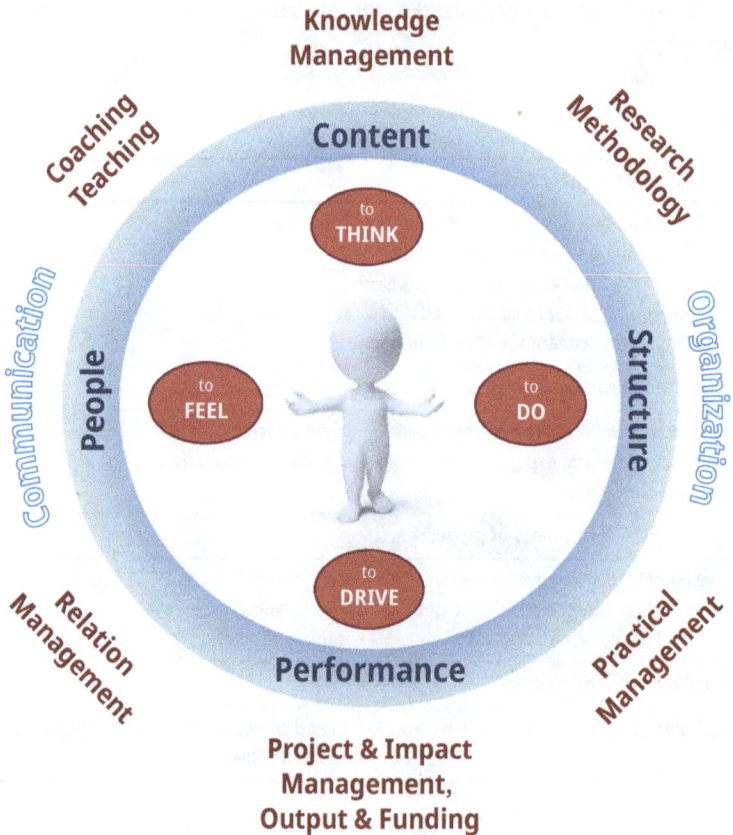

**Scientific Leadership:**

**creating an environment in which people flourish, grow, realize their own goals and contribute to those of the leader;**

**but ... it starts with self-leadership!**

Fig. 2.7: A summary of what scientific leadership entails.

Now let us try to come to a more formal description of leadership. Definitions have been given by Locke (1991): "leadership is the process of inducing others to take action toward a common goal"; and by Conger (1992): "leaders are individuals who establish direction for a working group of individuals, who gain commitment from these group members to this direction, and who then motivate these members to achieve the direction's outcomes."

The impact of effective leadership becomes clear in the following statement adapted from Harter et al. (2002): "Leaders impact the performance of their organizations through their immediate subordinates, whose influence spreads throughout the firm. The personality of the leader affects employee satisfaction, which in turn affects the performance of the organization." It is through motivating the people that the organization becomes successful.

In his 1991 book *The Essence of Leadership*, from which we took the definition above, Locke presents a leadership model based on four ingredients:

- **Motives** (drive and ambition to achieve and willingness to lead and use power to achieve goals) and **traits** (honesty, integrity, confidence, emotional stability, creativity, and flexibility)
- **Knowledge** (content, preferably from experience), **skills** (communication, handling conflict, and building relationships), and **abilities** (managing, planning, decision making, problem solving)
- **Vision** (development and seeking of commitment for it)
- **Implementation** (strategy, selection, and motivation of staff, team building, management)

In this view, leaders distinguish themselves from managers in the sense that leaders establish the vision, while managers implement it. Both qualities are needed; the leadership role can very well include management; the reverse is not true, however. The hallmark of leadership is having a vision, the inner drive to realize it, and the ability to motivate others to commit themselves to implementing the goals that derive from the vision:

> *Good leaders create a vision, articulate the vision, passionately own the vision, and relentlessly drive it to completion.* Jack Welch, CEO, General Electric

### 2.6.1 A working definition for academic leadership

In the context of academia, where education, discovery, and development bring in elements of idealism, visionary views of the future, and educating the next generations of leaders, we believe that the following description catches the essence of leadership:

1) Leaders have a vision, a dream of what they want to achieve. They are passionate about it. Their vision places the organization's goals in a larger context of, for example, the needs of the university, the scientific discipline, industry, or even the society. This vision is shared by all team members and serves as a source of inspiration and a common basis for a focused effort, without limiting

creativity and new ideas. Leaders are experts in their profession and keep a focus on long-term goals.

2) Leaders are great communicators and dispose over emotional intelligence. While cognitive intelligence refers to the objective logic involved in leadership, emotional intelligence is the ability to control one's own emotions and recognize those of others. Emotionally intelligent leaders succeed in creating a motivating/ stimulating environment and a social climate in which team members feel at ease to exploit their own strengths. Ideally, there are no barriers for critical discussion and constructive exchange of ideas. Cultural diversity is acknowledged and valued, and the leader and the more experienced group members function as mentors for the younger people. We discuss the concept of emotional intelligence further in Chapter 6.

3) Leaders know the essential management tasks needed to keep the organization running, such as keeping track of the progress of projects and the financial situation, maintaining a proper safety regime, ensuring that reports are ready in time, etc. Of course, they may have delegated many organizational and management duties to others, but they are periodically updated to be well informed about the actual position of the organization. We deal with some of the most important skills for academics in Chapters 3 and 4.

## 2.6.2 Some leadership styles

Thus far, we have defined the ideal leader as an all-round manager who masters his/ her discipline, has an inspiring vision of where to go, and possesses emotional intelligence. Does that apply to every leader? No, of course not. Rooke and Torbert (2005) presented an interesting classification of several types of leaders and their action logic, with their strengths and weaknesses, which are summarized in Tab. 2.2.

Tab. 2.2: Leaders and their main characteristics.

| Types | Characteristics | Strengths | Weaknesses |
|---|---|---|---|
| Alchemist (1%) | Generates social transformations. Reinvents organizations in historically significant ways | Leads society-wide change | None |
| Strategist (4%) | Generates organizational and personal change. Highly collaborative; weaves vision with pragmatic, timely initiatives; challenges existing assumptions | Generates transformation over short and long term | None |

**Tab. 2.2** (continued)

| Types | Characteristics | Strengths | Weaknesses |
|---|---|---|---|
| Individualist (10%) | Operates in unconventional ways. Ignores rules he/she regards as irrelevant | Effective in venture and consulting roles | Irritates by ignoring key organizational processes and people |
| Achiever (30%) | Meets strategic goals; promotes team work; juggles managerial duties and responds to outside to achieve goals | Well suited to managerial work | Inhibits thinking outside the box |
| Expert (38%) | Rules by logic and expertise. Uses hard data to gain consensus and buy-in | Good individual contributor | Lacks emotional intelligence and respect for the less experienced |
| Diplomat (12%) | Avoids conflict; wants to belong; obeys group norms; doesn't rock the boat | Supportive glue on teams | Can't provide painful feedback or make hard decisions |
| Opportunist (5%) | Wins any way possible; self-oriented; manipulative; "might makes right" | Emergencies and pursuing sales | Few people want to follow them long term |

Unfortunately, not every leader is, by definition, a pleasant person to work for. Toxic managers may display abusive, manipulative, and/or harmful behavior that can have detrimental effects on team members and the organization as a whole. They often prioritize their own interests over the well-being of their team and create a hostile work environment through bullying, micromanagement, favoritism, and lack of support or appreciation. The negative atmosphere caused by toxic managers can suppress creativity and initiative, as team members may feel demotivated or afraid to propose their ideas. Bosses may very well be tough and demanding as long as they are honest, respectful, and supportive of their staff (Daniel 2024).

**Leadership in action: the classical symphony orchestra under the maestro**
*Music ensembles present instructive examples of leadership situations. Let us take a symphony orchestra as an example. Our conductor, the maestro, is an accomplished musician in his late 50s. He has a **vision** that music is a universal language that unifies people across borders and races; he is passionate about the late romantic repertoire, and as a previous violin soloist, he knows many scores inside out. He prefers to work with young musicians, who he sees as the future generation of conductors to spread his philosophy over the world. The maestro's **mission** for the coming three years is to make a world tour with his young orchestra and to play in less privileged regions of the world to enthuse a large public for classical music.*

*In the intense rehearsals and concerts, the orchestra feels that the maestro stands above every symphony or solo concerto on the program, and the musicians are eager to play under his guidance and learn from him. The maestro leaves freedom to soloists in the interpretation of solos. During concerts, the orchestra is full of concentration and becomes ever more confident. Occasionally, something goes wrong, but then the conductor is the one who, with calm but clear gestures, gets everyone back on track. The joy of making music is visible and radiates over to the audience. This orchestra has almost become a self-organizing entity, and the maestro can fully concentrate on giving the music just that extra touch that makes it really special. Per-*

*formances are intense, sometimes exhausting, events where the audience generally feels enchanted. Musicians who played under the maestro are generally proud to mention this in their CVs.*

*The maestro and his orchestra demonstrate leadership in action, in combination with self-leadership of the orchestra, where different sections, e.g., strings, brass, wind, or percussion sections, function as optimal teams. The entire operation runs smoothly, although all concerts are different.*

## 2.7 Concluding remarks on leadership

In summary, successful leaders have a compelling vision that they share with their team, implying that they are experts in their subjects, they possess significant emotional intelligence, they are good in establishing and maintaining relations, and they have the necessary management skills, although they may have delegated many organizational duties to team members. We end this chapter with a paraphrased quote from Kathleen Schulz (2016), former director of the American Chemical Society, who in a column of *Chemical & Engineering News,* gave a concise description of leadership:

> *Great scientific leaders create a climate in which their team can do great work. Great leaders are passionate, take time to reflect, are always optimistic, are great communicators, and keep focus on long-term goals.*

We could not agree more!

The archetype of the alchemist and charismatic leader is Nelson Mandela (1918–2013), a leader of society-wide change in South Africa and admired and beloved all over the world, unparalleled in how he stood above the parties and in the end even respected and loved by his former enemies. We list some of his inspirational quotes:

**May your choices reflect your hopes, not your fears.**

**Do not judge me by my successes, judge me by how many times I fell down and got back up again.**

**It is better to lead from behind and to put others in front, especially when you celebrate victory when nice things occur. You take the front when there is danger. Then people appreciate your leadership.**

## References

Blanchard, K.H., Hawkins, L., Fowler, S.: Self Leadership and the One Minute Manager: Increasing Effectiveness through Situational Self Leadership; HarperCollins Publishers Inc., New York, 2005.
Conger, J.A.: Learning to Lead: The Art of Transforming Managers into Leaders; Jossey Bass, San Francisco, 1992.
Daniel, T.A.: Toxic Leaders and Tough Bosses; De Gruyter, Berlin, 2024.
Drucker, P.F.: Managing Oneself; Harvard Business Press, Boston, 2008.

Hay, L.L.: You Can Heal Your Life; HayHouse, New York, 1984.

Harter, J.K., Schmidt, F.L., Hayes, T.L.: Business-unit relationship between employee satisfaction, employee engagement and business outcomes: A meta-analysis. J Appl Psychol; 2002; 87; 268–279.

Locke, E.A.: The Essence of Leadership; Lexington Books, New York, 1991.

Manz, C.C.: Improving performance through self-leadership. Nat Prod Rev; 1983; 2; 288–297.

Manz, C.C.: Self-leadership: Toward an expanded theory of self-influence processes in organizations. Acad Manage Rev; 1986; 11; 586–600.

Rooke, D., Torbert, W.R.: Seven transformations of leadership. Harv Bus Rev; 2005; 83(3); 66–76.

Schulz, K.: Chemical & Engineering News, January 25, 2016.

# Guest column: Stand out by combining groundbreaking blue-skies research with collaborative industrial projects

Michael Claeys

*Michael has been a full professor at the University of Cape Town (UCT) since 2010 and is a leading international expert on synthetic fuels. For almost 20 years, he directed a national South African network (c\*change) that included students of all ethnic backgrounds at South African universities. It received international recognition for its inclusive education and research.*

In South Africa, students and academics are actively trying to bring about a transformative effort and ultimately create an inclusive and representative academic environment that offers equal opportunities for students from a diverse population. Government initiatives exist to support individuals from historically disadvantaged backgrounds to pursue higher education, followed by industrial or academic careers. Mentorship programs involving academics with motivational leadership capabilities are vital for guiding students throughout their educational journeys. By nurturing a diverse academic community, these endeavors aim to enrich the country with various perspectives and experiences, ultimately contributing to a more inclusive and equitable society and, not to forget, a rich pool of talent for the country's economy.

The country's challenges are many, but so are the opportunities. In my field of expertise, I mention the need for clean fuels for road, sea, and air transportation, which in South Africa are presently obtained from coal and natural gas or are imported. Eventually such fuels can only be sustainable if they are produced from $CO_2$, for which the technology is currently developed. The subject is an excellent source of opportunities for challenging research projects, requiring academia and industry to collaborate.

My advice to students aiming for an academic career is the following:

- **Stand out**: Develop unique ideas, approaches, or equipment to differentiate yourself. Become an expert in at least one field and avoid frequently switching to the latest trend. Demonstrate endurance in your pursuits.
- **Relevance of work**: While pursuing your ideas can lead to excellent scientific output and even intellectual property or business opportunities, remember that you and your work are not isolated. Collaboration with industry is crucial as it helps focus on relevant topics and ensures your research has real-world applications. Strive to balance industrially relevant research with exploratory, "blue skies" topics.
- **Teamwork and multidisciplinarity**: Emphasize the importance of teamwork and working across disciplines. Large, diverse teams, such as those in the c\*change initiative, showcase the value of varied backgrounds. Encourage self-leadership among team members.

- **Increase visibility**: There needs to be more than publications alone. Enhance your visibility through initiatives like organizing conferences (e.g., like we did with the triannual, international Syngas Convention conference series), developing unique equipment, or creating educational programs for schools to inspire high school students and teachers. This can attract prospective students to pursue academic education in chemistry, physics, or engineering and will eventually benefit you.

Building an academic research group requires well-developed leadership capabilities. This starts with practical management skills that make your life as a scientist easier. Then, you develop self-leadership and the leadership needed to guide and inspire the people working with you. I hope universities will widely introduce leadership training in their curricula and stimulate the active participation of all students, as it is an underestimated critical success factor for rewarding careers at all levels.

# 3 Presenting science: publications, talks, posters, and some ideas on conferences

## 3.1 Introduction: a crystal-clear message for a specific audience

Scientists and engineers need excellent communication skills to be successful. They regularly present their work, ideas, plans, and knowledge

- in spoken word at conferences during lectures, poster sessions, and discussion events, in seminars, group meetings, student courses, information events, laboratory tours, pitches during match-making events, interviews in the media, and
- in written text in articles, reports, lecture notes, posters, and not to forget grant applications, crucial for acquiring the funding for their activities.

Efficient communication (see the textbox) is a must-have skill for anyone who wants to accomplish something where other people are involved.

> **"Communication is an act of will directed toward a living entity that reacts."**
> - This crisp and clear definition of communication comes from Garcia, who bases his theory on the principles of war fighting. Let's analyze his definition further by using a paraphrased quote from Garcia's (2012) book, *The Power of Communication*:
>
> Communication is an act of will ...
>> Effective communication is intentional. It is goal-oriented. It is strategic. Unlike ineffective communication, effective communication is not impulsive or top-of-mind. Communication is not only about what one says, but about anything one does or is observed doing. It is about any engagement with an audience, be it a group of people or a single person, and it includes silence, inaction, and action.
>
> ... directed toward a living entity ...
>> Listeners are not passive bodies that absorb messages. Rather, they are living, breathing human beings. They have their own opinions, ideas, hopes, dreams, fears, prejudices, attention spans, and appetites for listening. Most important, it is a mistake to assume that audiences think and behave just as we do. Most do not. Understanding an audience and its preconceptions, and the barriers that might prevent an audience from accepting what one is saying, is a key part of effective communication.
>
> ... that reacts.
>> This is the element most lost on many leaders. The only reason to engage an audience is to change something, to provoke a reaction. Ineffective communication is not noticed, or it confuses, or it causes a different reaction than the one desired.

When presenting a story using PowerPoint slides, or explaining a poster, written and spoken words together form the most obvious content of what is conveyed. However, there are many more elements in a presentation that matter: think about intonation, facial expression, body language, the way a speaker responds to the audience, in answering questions or just in how he/she behaves at interruptions, e.g., when a mobile phone goes off.

https://doi.org/10.1515/9783111325644-003

In written text, e.g., articles, reports, applications, lecture notes, brochures, and web pages, each medium requires its own approach in terms of style, degree of formality, level of detail, balance between text and illustration, etc.

All this constitutes your side of the presentation, and you will be most successful if your presentation, report, or application is centered around a crystal-clear message that cannot be misunderstood.

The second factor is the receiving end of the communication, i.e., the audience or the readers: living entities that not simply absorb everything they hear or read (see the textbox on communication). Who are these people who are listening to you or are working themselves through your texts? What is their background? How do they take in and process information? How much time and effort are they willing to invest to get your point? How easily is their attention distracted? What causes this distraction? Is it you or your style of presenting or writing? Are there external causes, like construction activities in the next classroom, telephones ringing in the audience, two people whispering on the front row, or an inferior projector or sound system? Or in written text, is it the way you write, the layout of the text, the ineffective use or lack of figures and tables, that are maybe hard to understand?

Analyzing your audience or readership is very important, so that you can target your message at the right level. At the same time you should do everything to make your message as crisp and clear as you can and to present it as clearly as possible.

Our philosophy about presenting science is very simple:
1) Feel free to develop a style that fits you best.
2) Avoid the obvious mistakes that speakers and authors so frequently make.
3) Always build your speech, report, article, or poster around a crystal-clear message tailored to the people you want to reach.

## 3.2 Writing publications

*Manuscripts that address a relevant question and present a clear message based on original and new results, written in the correct format and in clear English, will always be published, but not always by your favorite journal.*

### 3.2.1 Why do scientists publish?

That sounds like a silly question, isn't it? Maybe you are a young PhD student who just reached an exciting conclusion of your first successful research project. Naturally, you can't wait to see your work published, preferably in a highly ranked journal that is widely read by your scientific community.

**Why scientists publish?**
**Scientific reasons**
- To present new results, methods, and insights
- To summarize the state of a field (perhaps from a specific perspective)
- To "claim" a subject (establish precedence)

**Strategic reasons**
- To acquire/maintain funding (both external and internal)
- To obtain status (personal, institutional)
- To qualify for a PhD degree, promotion, or tenure

Through publishing, you share your important findings with scientists all over the world. Simultaneously, you hope that you will become known for this success and that it may help you to build a successful career. Maybe your university requires publications before you can submit your PhD thesis. Your supervisor will obviously be pleased with the new insights gained, but he/she will also be happy that the research group can boast another good article on the publication list. It will be difficult without scientific production to acquire funding for the research. The university wants to see publications to prove its status as an important research institute. Hence, a successful publication is important in many ways, not only for its content per se.

But there are other good reasons for publishing your work. Writing down what you have done, and what you believe the results imply, is a very effective way to think deeply about the work's meaning, significance, and possible implications. The act of getting data, thoughts, descriptions, and interpretations on paper is quite confronting, and forces one to think deeply about the value of the work, its completeness, flaws in methodology, and validity of conclusions. This is a crucial phase in your research and a chance to substantially improve its quality, which you do not want to miss. Secondly, when a journal editor decides to consider your manuscript for publication and sends it out to experts for review, their reports may give invaluable feedback for revision, identification of weak spots and flaws in the argumentation, or offer alternative views on the significance of your work that you had not considered yet.

A study by one of the largest scientific publishers, Elsevier, conducted in 2005 revealed that authors still find the dissemination of new results the most important reason to publish. Still, factors such as furthering one's career, maintaining funding, gaining recognition, or establishing precedence as an expert on a topic were very important as a secondary motivation. The consequence is that authors and their institutions are not just satisfied to see their work in print, they also want to see it in a journal that has prestige in the specific field of the author or rather even in the entire scientific community.

### 3.2.2 Measures for prestige: Hirsch-index and impact factor

The scientific community has relatively recently (i.e., over the past 20 years) adopted two "quick and dirty" indicators for standing in the field, namely, the Hirsch-index (H-index) and the impact factor (IF). Neither of the two is normalized, e.g., for the age and experience of the author or for the size of the scientific domain or readership, implying that great care should be taken to interpret what these indicators mean. Both are based on citations, i.e., references to articles by other recognized publications. It is good to realize this limitation: parameters for the impact of scientific work are not based on real usage but solely on citations in scientific journals, which is not necessarily the same!

### 3.2.2.1 The H-index

The H-index is the number, H, of articles by an author (whether primary author or coauthor; there is no distinction in the roles that coauthors have played) that received at least H citations (Hirsch 2005). The way to determine someone's H-index is to rank his/her publications in decreasing order of citations and see at which number in the list the number of citations is less, as graphically illustrated in Fig. 3.1.

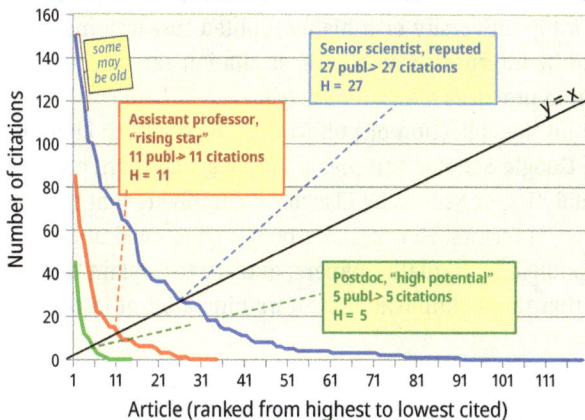

Fig. 3.1: The H-index is a commonly used indicator of the impact of an author's publications. H is the largest number of publications H that received at least H citations. Note that the H-index is skewed for age: the senior scientist's articles had a much longer period over which they could be cited than the postdoc, whose work is only recent.

To illustrate why the H-index is a "quick and dirty" parameter, one simply should note that an H-index of 10 implies that the author will have at least 10 papers and 100 citations in total, as a minimum. Suppose he/she has a colleague who also published at least 10 papers, with 5 of them hugely successful (50–100 citations per paper) and the other ones below 5 citations, his/her H-index is still 5. Nevertheless, the latter colleague has by far the highest scientific impact, in spite of a lower H-index. Another peculiar point is that any scientist who stops publishing after a few years will see the H-factor go up as long the work remains cited.

Nevertheless, the inventor of the H-index, Hirsch (2005), argued convincingly that the H-index has advantages over other indicators such as

- total number of papers: indicates productivity but does not reflect impact;
- total number of citations: measures the impact of one's work, but may be inflated by a few – perhaps atypical – successes, for example, coauthorship on a highly cited review article;
- citations per paper: allows comparison of scientists with different ages, but someone with an old successful paper continues to rank high after he/she stopped publishing.

Seniority remains an essential factor in judging the significance of someone's H-index. Hirsch compared publication statistics of colleagues in his own discipline (physics), and divided their H-index by the number of years since their first publication appeared, a parameter he called $m$. Hirsch suggested that an $m$-value of about 1 (e.g., an H-index of 10 after 10 years of scientific activity) represents a successful scientist, while $m=2$ (e.g., an H-index of 40 after 20 years) would indicate a truly outstanding scientist, likely to be found at a top university or a highly reputed research institute. However, such $m$-values should be interpreted with care, as citation rates depend on the size of a discipline and citation practices may vary as well.

Finally, H-indexes depend on the collection of publications from which citations are considered. In this sense, Google Scholar samples a larger space than, e.g., the Web of Science by Clarivate Analytics, or Scopus by Elsevier do, implying that the former usually gives slightly higher H-factors. Hence, don't be too concerned about the H-index, just do your best to publish high-quality papers, and try to explain the bureaucrats of your institutions that these simplistic indicators hide a lot of important nuance.

**Publication lists do not tell it all**

When starting his PhD program, Jean-Pierre joined an ongoing and successful research program under a famous supervisor. He could almost immediately start to carry out the ideas suggested by his supervisor, and experienced students showed him exactly how to use the equipment. He gathered data for seven good level publications together with his professor, who was happy with Jean-Pierre's performance. The papers were cited from the beginning, and when Jean-Pierre left, his H-factor was already 5 and on the rise. However, how much did Jean-Pierre really learn? Indeed, his next job, research scientist in an institute, was no success, as he had to design and build his own experimental setup. As interacting with other people was not one of his strengths, he did not get much done. His director terminated his contract after 1 year.

Javier was the first PhD student in the group in a new research program. He had to order and build up equipment and test it and was very happy that in the final year, he could do some successful experiments and write two publications. It took some time before these novel studies started to be noted and cited. His H-index is clearly lower than that of Jean-Pierre. Nevertheless, Javier is much better suited for a research career than Jean-Pierre is, who still has to demonstrate that he can build up an activity by himself. Good recruiters will recognize the difference, but unfortunately, in the present competitive climate, Javier may have more difficulty finding an academic position than Jean-Pierre, should they want to go in that direction.

### 3.2.2.2 The impact factor (and the citation half-life)

Prestige is based on how often your work is used by others, as measured by citations, also for journals. Publishers want a quick indication of success, and prospective authors want to know how well a specific journal is read, in the hope that their work has a better chance to be cited by others. Ironically enough, with all the search engines available these days, it is less important where a paper is published, as readers seldom read through hard copies of journals anymore. Nevertheless, the IF has become an indicator of prestige for journals, authors, and research institutions.

The IF of a journal is the total number of citations to articles in that journal in that year divided by the total number of articles published in the two previous years (Garfield 1972). Typical numbers are on the order of 1–5 for topical journals (i.e., of a specialized subdiscipline), 4–10 for general journals, 8–15 for reviews, and 20 or higher for prestigious journals like *Science* and *Nature* and some of their spinoffs.

There are several complications with the IF that users should be aware of (Moed 1996). First, it may be skewed by a few highly cited items, such as a review. Secondly, it is calculated by taking the number of citations to all published material, including letters to the editor, news items, comments, and obituaries, and dividing it by the number of scientific articles only. In addition, the IF concept does not account for the size of a field and how large the typical readership of the journal is (Waltman 2015). Another consequence of the definition is that as the scientific community grows and tools for online search and for swiftly including references in manuscripts become more widespread, many IFs tend to rise over the years.

Additionally, a period of minimum of 1 year and maximum of 2 years is very short for collecting citations, as a publication often has some induction time before it starts to be known and cited. To circumvent this, some journals define another IF on a 5-year basis; interestingly, these do not differ much from the usual 2-year values.

Sometimes, publishers try to manipulate IFs by artificially releasing issues at the beginning of the year (or prepublish papers in the Fall with a formal publication in January thereafter) or by systematically publishing a review of widespread interest in the first issue of a year. Some journals encourage authors to include at least three references to recent articles in the journal they are submitting to, which is a questionable practice, of course.

In this sense, the citation half-life is probably a more reliable indicator of a journal's quality, as a long half-life indicates that it publishes work of long-lasting value.

Hence, typical archival journals can boast citation half-lives of 10 years or more. The same is true for review journals (the well-established ones have both high IFs and long citation half-lives).

Although the journal impact factor is meant to compare how well journals are cited, it is often misused to assess the quality of individual articles, their authors, their laboratory, or even their institutions. Why is a parameter that measures how soon and often the articles of a journal are cited perceived as a mark of quality for authors? Because a high IF is regarded as an indication of prestige for a journal, and hence, getting accepted in such a journal is regarded as an accomplishment. Nowadays, we often see that applicants for academic positions include impact for the journals in which they published in their CV, which is a questionable practice. Again, experienced scientists rather focus on content, but recruiters are often fond of simple indicators.

In conclusion, prestige indicators such as H-index and IF are easy-to-calculate benchmarks based on the number of citations in recognized scientific journals, but one must be aware that they disguise a lot of detail and therefore should be used with great care.

### 3.2.3  The basic principles: a clear message for the readers

Scientific writing follows simple principles: the author has a clear message and he/she wants it to be read. Readers want a crystal-clear and easy-to-digest message about an issue in which they are or may become interested. To this end, articles follow a more or less standardized format, in which readers know where they can find what they want to know.

In an era where publications have almost become scientific currency" and the fact of having them in the personal or institutional portfolio is often as important as their content, it is good to realize why we do publish. In his book *How to Write a Paper*, Richard Smith (1994) proposes the following questions to prospective authors who are about to write a manuscript:

1) What do I have to say?
2) Is it worth saying?
3) What is the right format for my message?
4) What is the audience for my message?
5) What is the right journal for my message?

Journals and their readership expect a focus on quality and novelty. In Tab. 3.1, we quote from our distinguished colleague Professor Roel Prins (ETH Zurich; long-time editor of the *Journal of Catalysis*), who taught many courses on scientific publishing:

Tab. 3.1: What quality journals are looking for (courtesy: Roel Prins, editor).

| Wanted | Not wanted |
| --- | --- |
| Originality | Duplication ("me too research") |
| Significant advances | Reports of no scientific interest |
| Appropriate methods and conclusions | Work out of date |
| Readability | Inappropriate methods/conclusions |
| Studies that meet ethical standards | Studies with insufficient or incomplete data |

Another relevant quote for strategic publications, e.g., those of authors seeking to fill their publication portfolios rather than making a novel contribution, is the following:

*Just because it has not been done before is no justification for doing it now*
(Peter Attiwill, editor-in-chief, *Forest Ecology and Management*)

If you are convinced that you have something important to publish, you can go ahead and start preparations. But first it is good to review once again what scientific writing is about (see Tab. 3.2). We will now go through the stages in preparing a manuscript.

Tab. 3.2: The elements of scientific writing: clarity, objectivity, accuracy, and brevity.

| Be clear: | Be objective: |
| --- | --- |
| – Ordinary, short, familiar, nontechnical terms are better than long, grand, unfamiliar technical and abstract vocabulary.<br>– Avoid complicated sentences.<br>– Realize that English is a foreign language for many authors and readers. | – Be unbiased, unemotional, and truthful.<br>– Give evidence to support your arguments, and acknowledge merit in other views.<br>– Separate results from interpretation.<br>– Benchmark results against the best available information in the literature. |
| **Be accurate:** | **Be brief:** |
| – Facts need to be accurate, complete, and verified more than once.<br>– Avoid ambiguous or misleading statements.<br>– Carefully check all data and procedures that are reported.<br>– Include indications of accuracy and reliability, such as error margins, standard deviations, etc. | – Use words efficiently but avoid uncommon abbreviations and jargon.<br>– Give a concise "Introduction."<br>– Give a brief discussion.<br>– Provide effective, self-explanatory illustrations, with informative captions, and clear tables.<br>*A high-quality picture is worth a thousand words.* |

Jilin, a PhD student at an Asian university, had little success with her research so far. She could, for her thesis, only write chapters on introducing her subject and on the experimental procedures she had used in her research. Together with her supervisor, she decided to select a publication from the literature and to repeat the work with a slightly altered composition of the central material – a catalyst – but keeping the essential ingredients the same. The "me-too approach" appeared successful and Jilin obtained slightly different results as in the paper she used as inspiration, although not so that conclusions would have to be changed. Jilin wrote a thorough paper, which she discussed at length with her supervisor, and she confidently submitted it to a well-established journal of the field. However, the editor returned it after a few days. He found the paper technically sound but did not feel that he learned something he did not know yet. Jilin was disappointed but submitted it to another journal, where she initially had more luck. The editor sent it to three reviewers. One recommended publication after some revision, the second had some doubts about novelty and advised to shorten the manuscript substantially, while the third reviewer happened to be the author of the original paper Jilin based her work on. He was not impressed at all and recommended rejection due to lack of originality. The editor had little other choice than to refuse the paper. Finally, Jilin managed to get the paper published in a new open-access journal that was not so critical in its selection procedures yet. So, in the end, Jilin had her published article for inclusion in her PhD thesis, but not at the level she had hoped for. Of course, we sympathize with Jilin and are happy that she succeeded to graduate. However, we strongly disagree with university policies to require published articles before a student can graduate. In the case of Jilin, she clearly demonstrated that she can perform state-of-the-art research at the international level. That it was not at a particularly high degree of novelty is something the supervisor should be concerned about. We find it an inappropriate requirement for a young student, whose main interest is to learn from the experience.

### 3.2.4 Stage 1 – before you write

### 3.2.4.1 Study the Guide for Authors

When you know what you want to publish, and where, then our first recommendation is to read the Guide for Authors of the journal you have in mind. You would be surprised to learn how many inappropriate manuscripts journals receive. These are mostly rejected right away simply because authors failed to check the scope of the journal and the requirements for submitting a manuscript. Also, should the journal of first choice reject your paper, then never submit the same material in the same form to a second journal without checking its Guide for Authors first!

### 3.2.4.2 The message in one sentence

Try to capture the essence of your paper in a single sentence. This sounds simple, but it is not. In a sense, it is the analogue of presenting your message in 30 s, sometimes referred to as the "elevator pitch" (see Chapter 4). However, if you manage to catch the intention and conclusion of your work so concisely in one sentence, it means that you know exactly what the focus of the paper should be and what needs to be included.

An example for such a one-sentence message (basis for a paper in the *Journal of Physical Chemistry*, we apologize for the specialized nature of the content) is as follows:

*Polyethylene formation from ethylene with a chromium catalyst occurs on atomically dispersed, divalent chromium ions coordinated to the silicon oxide support via two oxygen ions, as evidenced by data from surface spectroscopy and electron microscopy obtained on a model system for the industrial catalyst.*

Such a "message in one sentence" is your guide to select (i) topics for the introduction, (ii) results, (iii) explanations, (iv) description of methods, and (v) pertinent literature that you need to include for a truly convincing story.

### 3.2.4.3 Cite the pertinent literature

**Literature from before 2000 is not necessarily outdated**! Young academics sometimes seem to think that science – or at least the relevant stages in their discipline – started one or two decades ago while everything discovered before 2000 is textbook material. Much excellent research and thorough thinking was done throughout the previous century. Methods of investigation and computer power to interpret and simulate measurements may be much more refined nowadays, but many scientists in the previous centuries had a broad and thorough education in the basic disciplines and sometimes very clever ideas. In addition, they were under much less pressure to acquire funding and turn out papers. Hence, the articles they did publish were often meticulously prepared and very well-thought through. In chemistry and physics, many articles and books from the early 1900s and onward offer exciting ideas and can be a great source of inspiration. Do yourself a favor and take a few hours now and then to browse through the old monographs and journals of your field (hopefully, your library still has them, but you can find some famous older literature online as well).

Next point on the agenda is to review again all related literature and to identify the key papers that must be included in your manuscript. Be careful here; it is tempting to collect a long list of references from a service such as Scopus, Google Scholar, or the Web of Science and to include all these in your manuscript. However, try to be selective and cite only the key papers that you have read or still must read before you finish your manuscript. Keep in mind that it is an ethical obligation of every author to give proper credit to publications you have used in any way, particularly in the design of experiments, the interpretation of data, and the construction of theories in the discussion. It is also a moral obligation to check the literature carefully for papers on your subject, and you may have to go back into the previous century (see Textbox). On the other hand, it is not considered proper to cite for purely political or strategic reasons, for example, to help a befriended colleague or to please a person who might be the referee on a future grant proposal. Excessive self-citations fall in the same category, but of course, it is acceptable, and in fact expected, that you refer to your previous work on the subject if it is related to your present manuscript. References should be given in the proper format as requested by the journal, taking care that all names are spelled correctly.

By the way, citations and references are not the same: citations are short indicators like "Author et al. [1]" or just [1], which one includes in the text, while references appear at the end and contain the full bibliographic details of the citation.

**Strategic citations – not always the author is to blame.**
A questionable example of strategic citations is the following. A recent manuscript from our own laboratory was reviewed for a well-established journal and found worthy of publication after some revision, provided – and this was a request from the editor! – we included more citations to recent work published in his journal. Apparently, the editor found this an effective way to boost the impact of his journal. Such a request is, of course, unethical.

### 3.2.4.4 Who qualifies for authorship?

Authorship represents the formal acknowledgment of a scholarly contribution to a published or presented piece of information, be it a paper, patent, conference abstract, or presentation. Who qualifies for this explicit status and who not can be a tough question, and recommendations for authorship from learned societies and publishers vary. The issue is not so much who contributes to the actual writing, but what other contributions are substantial enough to be acknowledged by coauthorship. Points to consider:

– who envisaged the research idea (and placed it in the proper context),
– who organized the funding,
– who devised a detailed research plan,
– who furnished the infrastructure and know-how for the experiments,
– who carried out experiments, computer simulations – in general who acquired data,
– who analyzed and interpreted data,
– who assessed the new knowledge and placed it in context current scholarly thinking on the subject,
– who wrote the manuscript, or contributed to it,
– who takes responsibility for the final version?

The next question is how large a contribution should be to qualify as "substantial."

You – and if you are a PhD student or postdoc, together with your supervisor – will have to decide who should be invited as coauthor on the paper. Consider that modern science is teamwork, and several people may have given essential input for the research that is the basis of your manuscript. What about indirect but vital contributions, such as the laboratory infrastructure in which you have worked and the funding for the work; should these be considered as justification for participation in authorship? Different organizations have different rules, and your supervisor or director will probably have a policy for handling such issues. Many journals ask for clarification of the roles of each individual author when you submit the manuscript, but be aware that policies for coauthorship are not uniform.

The order in which the authors appear also requires serious consideration. There are two favored positions: first and last author. Sometimes two persons (e.g., two PhD students) made equal contributions to a study, such that both could be considered for the first author position. In such cases, an asterisk on their names and a footnote can be used to indicate shared first authorship. Some organizations value the status of corresponding author; journals often indicate this by an asterisk or a footnote. Who deserves the preferred positions? Usually, the person who did most of the work and wrote most of the paper is the first author, while the most responsible author – e.g., the supervisor or sometimes the director – takes the last position. However, some laboratories practice different rules, for example, to always place the most senior professor's name first or last or to put names in alphabetical order. In our laboratory, the student or scientist who does most and writes the manuscript is the first author, while the senior scientist who takes final responsibility is the last author. Hence, in our publications, you will find the director's name often in the middle, as the younger staff is usually responsible for the project and for daily supervision of the students involved and takes final responsibility for the content of the paper (and serves as the corresponding author). Whether prestigious or not – for us, it is merely an administrative role.

### 3.2.4.5 Which journal to choose?

An estimate suggests that in 2014, there were approximately 35,000 peer-reviewed academic journals (Scott, 2017). Many authors nowadays use the IF as a guide for selecting journals, or they select one of the well-established journals of their field. Some authors will try to publish as much as possible in a more general journal of high reputation. If rejected, they go to the next one in the – sometimes only perceived – reputation list, until the paper gets accepted. Be aware, however, that most journals have a clearly defined scope, as described in the Guide for Authors, and that the first thing an editor will do is to judge whether your manuscript fits in. You may use the list of references of your manuscript as a criterion to judge if your choice of journal makes sense. If your target journal, or related journals, do not appear in your list, you may have made the wrong choice. Editors often also use this criterion as a quick indicator for appropriateness.

We repeat: whatever journal you choose for submitting your work, make sure that you always read the Guide for Authors. Your manuscript must satisfy the requirements for, for example, layout, section lengths, nomenclature, number and type of figures and tables, and reference style. Do not annoy the editor by neglecting either the scope of the journal or its requirements for layout.

**Predatory journals:** Beware of publishers whose priority is to earn money, rather than disseminate and archive trustworthy new scientific knowledge. Predatory publishers charge high fees to authors, but provide minimal editorial support, including superficial peer reviewing, if any. Authors feeling under pressure to publish – e.g., to brush up their CV – may be seduced to submit work to such journals, but we strongly advise avoiding journals with questionable procedures at all times.

### 3.2.4.6 Ethical obligations of authors

Unfortunately, the world of scientific publishing has seen quite a few cases of miscon- duct recently, ranging from questionable practices such as duplicate publication, self- plagiarism, or violations of good practice by not revealing complete sample informa- tion on materials or methods, to outright fraud, such as falsification and plagiarism. The following list (Dodd 1997) summarizes the main ethical principles:

- Present an accurate account of your research and an objective discussion.
- Provide sufficient detail and references to enable repetition of the research.
- Identify hazards.
- Cite the pertinent literature and avoid excessive self-citation; identify the source of all information quoted, except what would generally be considered as common knowledge.
- No unnecessary fragmentation of research reports.
- Inform the editor of related publications under consideration elsewhere.
- No duplication of previously reported work (except in a review).
- Criticism of other work is permitted, but never criticize person(s).
- Coauthors share responsibility and accountability for the contents.
- Mention sponsoring parties and disclose any possible conflict of interest.

True scientific leaders will not only adhere strictly to ethical principles themselves but also teach these actively to their students and coworkers.

### 3.2.5 Stage 2 – writing the paper

#### 3.2.5.1 Work from a basic structure; preferably separate results and discussion

After you are clear about the message of your article, and you know what its format will be, comes the moment to sit down and start writing. It is good to have a basic idea of what the eventual length is going to be. Table 3.3 lists the well-known basic report structure that most journals adhere to for regular articles (letters and short communications usually have less or no sections). We have also included suggestions for the length of the manuscript as a start. Of course, you use as many pages as it takes, but try to keep the manuscript concise, and do not write overly long introduc- tions and discussions. In addition, keep the number of references to what is needed, and don't inflate the list (see above: cite the pertinent literature). Some journals re- quire you to use their template for the submission of a manuscript.

Nowadays, the trend is toward shorter papers with more material in a "Supple- mentary" section. However, the risk is that many readers will never look at this sec- tion, which implies that material you place here may essentially go lost. Therefore, we recommend keeping this section to a minimum and avoid using it to publish impor- tant results that would be valuable for their own sake.

Unfortunately, another trend is to publish articles with a combined "Results and Discussion" section. Our strong recommendation is to separate these parts. Often the main value of your paper is in the novel data that you are reporting. If documented properly, these results are of long lasting value. However, you will be interpreting their meaning in a context of what is known *now*. Insights usually change over time. If your results are presented within a framework of thinking that may be obsolete in the future, they will probably not be used anymore. Results, provided they are presented well and properly documented with the conditions under which they have been obtained, can have eternal value; interpretations may be outdated tomorrow.

In the same vein, accurate and complete descriptions of the materials used and the experimental procedures employed are essential for the value of your article. It is an ethical obligation of authors to present the work such that peers in the same discipline can repeat the work. Hence, leaving out crucial details is not allowed, although it happens unfortunately, e.g., to protect competitive advantages.

We strongly recommend you refrain from superlative writing, i.e., using adjectives such as "novel," "excellent," "remarkable," "extraordinary," "high-quality," or "state-of-the-art," all of which are essentially meaningless. Additionally, words like "robust," "novel," "innovative," or "unprecedented" add no quantifiable significance to your words and may even make experienced readers suspicious (Scott 2017).

**Tab. 3.3:** The structure of articles and reports.

| Sections and suggested initial length (manuscript pages) | |
| --- | --- |
| Abstract | 0.5 |
| Introduction | 1–2 |
| Experimental methods and materials | 1–2 |
| Results | 3–6 |
| Discussion | 2–4 |
| Conclusion | 0.5 |
| Acknowledgments | As needed |
| References | As needed |
| Supplementary material | As needed |

### 3.2.5.2 What to do about writer's block

You started writing, and you finished the first lines, which took longer than expected. Even worse, the next line even takes more effort to get it on the screen. Although you formulated your 'message in a sentence" (see above) and started with a clear idea about what needs to be written up, the right words don't seem to come. Every sentence you write looks bad, inaccurate, poorly formulated, and does not quite express what you wanted to say. This happens often.

Chances are high that you are suffering from writer's block. It is a common phenomenon, particularly among less experienced authors. The basic reason behind writer's block is that you are trying to achieve two things at the same time, which by nature don't go along too well:

- Getting your ideas on paper is an act of creativity; it uses the right side of the brain, where intuition, creativity, and appreciation for art and music reside.
- Formulating clear and correct sentences, however, is a skill-based effort; it uses the left side of the brain, where analytic capabilities, logic, language, scientific, and mathematical skills reside.

Combining creative ideas with technical skills to express an idea in a perfect sentence implies that you block one activity by the other. Hence, learn to apply the following principle:

*Write first – then get it right*

In other words, write down what comes to mind, and don't worry about language, correct grammar, sentence structure, or spelling. Tomorrow, you can revise your text and get it in proper shape.

Another piece of advice that will help you is to feel free to start in any section where you know what to write or even to add to different sections in random order. Hence, write in a convenient order. Here, the recommendation is also: write first and then get it right. See Tab. 3.4 for a suggestion, but do it as it works best for you.

An old but highly appreciated guide on writing correct sentences is the famous little book *The Elements of Style*, originally written in 1918 by William Strunk and later expanded by White. It is still regarded as one of the most influential texts on writing in English (Strunk 2009). Nonnative English-speaking authors may appreciate a helpful book by Glasman-Deal (2010).

Tab. 3.4: Writing the manuscript.

| Write in a convenient order (and feel free to do it your way) | |
| --- | --- |
| **Experimental** | Brief but complete |
| **Results** | Clear figures, informative captions, complete tables, and describe them |
| **Discussion** | Summarize key results, asses their value, place in context, discuss significance |
| **Conclusion** | In essence, an elaborate version of your "message in a sentence": Start with the aim of the work, then list the conclusions, followed by one or more sentences of perspective on what next |
| **Introduction** | Context, aim of your work, key literature, identification of a clear research question and your (novel) approach, preview of the message |
| **References** | Use reference software; limit references to ones that are needed |
| **Title** | Short form of your message, specific, concise, inviting to read |
| **Abstract** | What, why, how, and significance, clear message |

### 3.2.5.3 Attractive illustrations and tables with understandable captions tell a story

Figures are extremely important and should be laid out with care. Often, however, the figures we see in publications are far from ideal. Please realize that the common software for presentations and spreadsheets *do not at all* produce good figures – these are only a starting point you want to improve upon. Place carefully designed texts on ordinates, insert labels on spectra and curves inside the figure and not in legends or captions, and use labels that are understandable and not some type of secret code that only the author understands (see Fig. 3.2). Figure captions should be clear, informative, and attractive, such that the figure can be understood without reference to the text.

The same principles hold for tables; also, these should be able to stand on their own and tell a part of the story. Every author can make high-quality illustrations, but it takes effort (which is well invested!). Just realize that several readers will not have time to spell out the text, but rely on title, abstract, figures and captions, and the conclusion before they may decide to read the entire paper. Initially, even more important, the editor to whom you submit may do the same.

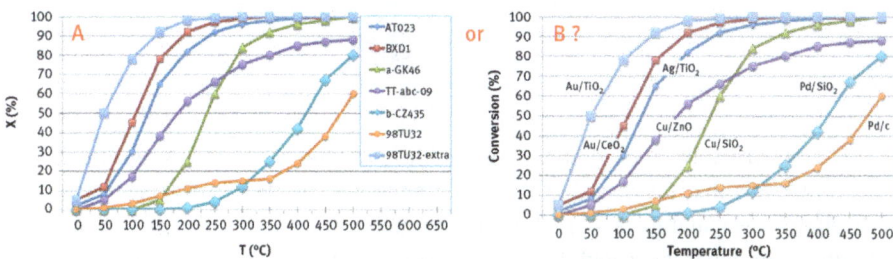

Fig. 3.2: Two examples of how data can be presented: (A) as taken directly from a spreadsheet with measured data, along with experiment codes from the laboratory and (B) the same graph but with understandable labels on axes and data. If you present graph A to a reader, he/she will mostly be busy with looking back and forth between the curves and the codes in the legend while trying to remember what these secret codes stand for. If you show the right version, the reader can immediately concentrate on the meaning of the data, as the figure largely explains itself (the graph shows the degree of conversion of a certain chemical reaction in combination with different metal catalysts on support materials).

### 3.2.5.4 Sections that often get too little attention: title, abstract, and keywords

It sounds strange, but many authors do not spend enough effort on these elements that will be freely available to everyone. Nevertheless, title, abstract, and keywords are the first elements on which editors decide if they will consider the manuscript, while later prospective readers use it to decide if they want to see the entire paper.

The title is the main attention getter; it should be brief, specific, and attractive and contain signal words that trigger interest. Constructions such as "A study into the effect of . . ." essentially waste space and look boring. Think back to the "message in a sentence" – maybe a shortened version could serve as the title? Instead of listing examples, we strongly advise you to browse through some recent issues of your favorite journals and try to identify some effective titles.

Keywords are important for abstracting services and for internet search machines and can be essential for prospective readers to find your paper; hence, providing a careful selection of such terms is recommended. Often, "key phrases," which combine a few words, will reflect content more accurately, and many journals allow these.

The abstract is of decisive importance for getting your paper published and read. Editors use it to judge whether your paper fits in the scope of the journal and readers to decide if they will read your paper. Use the abstract to convey your message, and don't use it as a verbose list of contents. An efficient abstract is written around the message of your paper; it explains the significance of the work, motivates why the research has been done, and conveys the major conclusions. Good abstracts are brief, specific, and accurate; avoid hype; use no abbreviations or technical jargon; and cite no references.

Title, keywords, and abstract, together with author names and affiliations, form content that appears freely online – visible for everyone, and not only for those with access to the journal. Hence, give these crucial elements the attention they deserve.

### 3.2.6 Stage 3 – submission of your manuscript

Manuscripts are submitted online. Many publishers have perfected their submission portals to levels that the actual process of submission can be a breeze, provided you are well prepared and have all requested documents ready. Carefully check the Guide for Authors on what is needed. Incomplete submissions are generally not considered and returned to the author.

#### 3.2.6.1 Convince the editor to consider your paper: the submission letter

Write a clear and concise submission letter to the editor. Your aim is to convince the editor that your paper is worthy of consideration for the journal and that it should be sent out for peer review. Check if the Guide for Authors gives any requirements on submission letters and then carefully draft it, addressing in any case the following points:

- The message of your paper and why this message is important, novel, and deserves publication.
- A motivation for choosing this journal; praising the editor for a high IF may not be convincing – referring to the scope of the journal and listing a few papers on related subjects that the journal published might be more effective.
- If you have coauthors, provide a statement that all agree with the content and have given their permission to submit the work. Often, a short explanation of each author's role is requested by the journal.

- A statement that the work has not been published, even not in part, and is presently not considered by any other journal. If you have submitted a related manuscript elsewhere, you must mention this and include a copy for the editor's information, so that he/she can verify that there is no overlap or that the main conclusions have not yet been given away.
- A few suggestions for reviewers who can give an objective expert opinion on your manuscript. Avoid mentioning colleagues from your own institution or coauthors from your recent papers, and it is also not helpful to suggest the leading star scientists in the field. The editor may invite one of the people you suggest, but more likely will he/she add these names to his/her database for future use. It is acceptable and even advisable to indicate individuals who should *not* be invited for review due to conflicts of interest.

### 3.2.7 Stage 4 – the editorial process: editors and reviewers

The editor is the intermediate between author, publisher, and the scientific community, a sort of "broker" who understands the interests of all parties involved. Associate or assistant editors can assist the editor in chief. For quality journals, editors are almost always experts with considerable seniority and recognition in the international community of their field. They have extensively published themselves and reviewed many articles. Their stature in the scientific discipline gives them the authority needed to take important decisions:

- on the scope of the journal (usually together with the editorial board of equally well qualified experts from the field);
- on the manuscripts that have been submitted to their journals; and
- on matters of ethics and scientific misconduct.

Figure 3.3 gives an impression of the timing in the handling of manuscript, taken from the work flow of a typical, well-established topical scientific journal publishing several hundred papers per year and with an IF in the range of 5–10.

#### 3.2.7.1 Your paper receives a first decision

Normally, a support office at the publisher's assists the editors. Often it serves several journals. When your paper is received, the office will log it in the system, check if the submission is complete, ask you for additional information if needed, and finally will forward it to the editor.

The editor then applies a first screening, implying he/she briefly assesses the subject, the quality, and the novelty of the manuscript and he/she decides whether it fits in the journal's scope and whether it meets the requirements for quality of presenta-

## Duration of the Editorial Process

1000 Submitted manuscripts; Average handling time in days;
Rev = Review process; Au = Author(s); Off = administrative support office

| | Submission + Editorial screening | Handling Editor | Rev | Handling Editor | Au | Off | Handling Editor | Rev | Handling Editor | Au | Off | Handling Editor | Rev | Handling Editor |
|---|---|---|---|---|---|---|---|---|---|---|---|---|---|---|
| **Days** | 3 | 3 | 30 | 2 | 24 | 2 | 2 | 23 | 2 | 12 | 1 | 4 | 18 | 2 |
| Min-max | 0-10 | 0-21 | 2-60 | 0-21 | 0-50 | 0-5 | 0-21 | 0-60 | 0-12 | 1-28 | 0-5 | 0-21 | 3-30 | 0-4 |
| **No of Ms** | 477 admitted / 108 transferred / 415 rejected | 452 admitted / 5 transferred / 20 rejected | 452 | 261 to revise / 11 withdrawn / 180 rejected | 261 | 261 | 80 review / 181 accept | 80 | 52 revise / 28 accept | 52 | 52 | 11 review / 41 accept | 11 | 11 accept |
| **Cumul. days** | | 6 | | 38 | | | 66 | | 91 | 103 | | 109 | | 128 |
| Ideal days | 2 | 1 | 21 | 2 | 14 | 1 | 1 | 7 | 2 | 4 | 1 | 1 | 4 | 1 |
| Ideal total | | 3 | | 26 | | | 42 | | 51 | 55 | | 57 | | 62 |

| 548 rejected/transferred within 1 week (0.1–3 weeks) | 180 rejected after 5.5 weeks (1-18 weeks) Total Out: 739 | 181 accepted in 9 weeks (4-16 weeks) Total Accepts: 181 | 28 Accepted in 13 weeks (7-20 weeks) Total Accepts: 209 | 41 Accepted in 16 weeks (10-22 weeks) Total Accepts: 250 | 11 Accepted in 18 weeks (12-22 weeks) Total Accepts: 261 |
|---|---|---|---|---|---|

**Fig. 3.3:** Example of the work flow for a typical scientific journal (in days from the date of submission).

tion. If positive, the editor may then decide to delegate the manuscript to an associate editor for further handling or to send it out for review immediately.

**Single-blind peer review is the norm**

Different systems of peer review exist in theory:

– Open review: when names of reviewers are revealed to authors (and sometimes readers)
– Single-blind review: reviewer names known only to the editor and not to the authors
– Double-blind review: names of reviewers and authors are unknown to both sides.

The double-blind system is often considered as the most ideal, as any prejudice on the authors or their previous work would be excluded. However, in practice, one can almost always recognize at least the main author, e.g., from citations made in the introduction. Therefore, the single-blind system is the common mode of peer review. For open review, the fear exists that many invitations for reports could be turned down because scientists are not keen on being identified as the one who recommended negatively (which could have implications when roles are reversed in the future).

If his/her decision is negative, the editor returns the manuscript to the authors, along with an explanation why he/she finds the work unsuitable for consideration. While you as an author will be disappointed, you should also realize that the editor has a good feeling for which manuscripts have a chance to get positive recommendations in the review process. By his/her off-hand rejection, you have a quick decision and can prepare for submission elsewhere, while the editor also ensures that he/she does not unnecessarily invoke reviewers, whose time is precious as well.

### 3.2.7.2 Peer review

Manuscripts found worthy of consideration are sent out to peer reviewers, usually two to four, who are asked to critically assess the manuscript and to report back in a few weeks.

Reviewers are specialists from the field, often well-established scientists, but sometimes also younger postdocs or even PhD students at the end of their projects, who read the manuscript and then give their opinion on the quality and novelty of the manuscript, its suitability for the journal, and they often indicate where corrections are needed.

Peer review is a system for quality control and improvement, from which the entire scientific community benefits. Note that reviewers do not decide on acceptance or rejection, which is the role of the editor.

> **Reviewing is a learning opportunity**
> Serving as a reviewer for manuscripts, conference contributions, or grant proposals is a wonderful opportunity to learn. It gives you insight in how your peers do their research, how they think, and how they plan their next research.

As the quality of scientific publishing depends critically on the mechanism of peer review, every scientist who is or will be an author has the obligation to do a fair amount of reviewing. Writing a balanced report takes time, but it is also a learning experience, and many scientists regard it as a privilege that they have to study certain manuscripts in depth. We therefore recommend young scientists, i.e., PhD students in their last year and postdocs, to become engaged, for example, by asking their supervisors if they can assist and write a report together. This is an excellent opportunity to learn.

Peer reviewers are bound to ethical obligations; we list the most important hereafter (Dodd 1997). Reviewers should

– accept only invitations for manuscripts inside their own area of expertise;
– judge objectively and respect the intellectual independence of authors;
– explain their judgments, if possible supported by references from the published literature, and refrain from unsupported assertions;
– be alert to correct and complete citation of relevant work;
– treat manuscripts fully confidentially, be sensitive to conflicts of interest, and not use information from unpublished manuscripts in their own work, unless with consent from the authors;
– refuse to assess papers from authors who are too close, such as friends, collaborators, or former supervisors; and
– return their report within 2–3 weeks at most.

### 3.2.7.3 Editorial decisions based on peer review

When all reviews are in, the editor takes a decision on the basis of the reports, which can be any of the following:

- acceptance (direct acceptance is rare, however),
- acceptance on condition that the authors make minor but mandatory revisions (if they do not accept these, they should provide a convincing reason for this),
- revision required before a final decision can be taken – implying that the revised paper will be reviewed again by the editor or the original reviewer(s), and
- rejection.

If reviewer recommendations conflict each other, the editor may invite additional reviewers, e.g., from the editorial board, or he/she may rely on his/her own judgment.

In case a paper is rejected, be it in first or second instance, authors always have the right of rebuttal. A request hereto can be directed either to the handling editor or to the editor in chief. Usually, editors will respond positively to an appeal from the authors, and they will either provide further clarification of their decision or they may decide to request another expert opinion, often from an editorial board member. However, a second decision is final.

Rejection is of course a disappointing outcome, but it happens often, even to well-established researchers and even to world-famous Nobel Prize winners. The important point is to not take such decisions personally but to evaluate honestly why the paper was rejected. Are your conclusions truly warranted? Would inclusion of other or more results enhance the chances of acceptance? You always have the right of rebuttal, and sometimes, one must fight for a paper, but often, choosing another journal is the best option.

### 3.2.7.4 Dealing with revisions

In case the decision is that your paper might qualify after appropriate revision, the author should consider all comments carefully and revise the paper where it is necessary. He/she should also write a reply to the reviewers and indicate the actions that have been taken or explain convincingly why the manuscript was not changed. The experience of many editors is that authors can be quite slow in submitting revised manuscripts; this is often a step causing unnecessary delays in the publication process. A period of 2–3 weeks is acceptable, but generally, it is best to deal with revision as soon as possible.

After submission of the revised paper, the editor may take a decision by himself/herself or send the paper out for review again. Let's hope he/she decides to accept your work now. If not, learn from the comments and submit an improved version elsewhere. If the work has merit, it will certainly get it published!

### 3.2.8  Stage 5 – promote your work

Congratulations! Your paper has been published. Next, you will ensure it to be read, used, and cited. You should be aware that the average number of citations is on the order of one per paper per year only, and depending on the scientific discipline. Hence, make sure that people know your work. You may send PDF reprints to selected colleagues in the field, present your work at conferences to promote it, and put an abstract and your best figures on the web, e.g., on social platforms such as ResearchGate or LinkedIn. In any case, promotion of your work is important and citations do not usually come automatically.

### 3.2.9  In conclusion

How do you get your paper accepted? Focus on a clear message based on original results that address a relevant problem or question; stress why you did the work and what comes out; and ensure that you write the paper in the proper format and correct English. Never forget to check the Guide for Authors. Success!

## 3.3  How to give successful presentations

How often have you been listening to oral presentations about interesting science while you nevertheless had difficulty to pay attention until the end? How often did you lose interest or got distracted before the speaker had even come halfway? Was it the subject and content of the talk, or was it the way the speaker presented it?

Many presentations deal with interesting work but are difficult to follow because the speaker unknowingly makes several presentation errors. By far the largest mistake is that many speakers do not realize how audiences listen. Go back to the beginning of this chapter and read how Garcia describes an audience as *"living human beings with their own opinions, ideas, hopes, dreams, fears, prejudices, attention spans, and appetites for listening . . . it is a mistake to assume that audiences think and behave just as we do."*

Our advice for giving effective presentations is simple: feel free to develop your own style, but do your best to avoid the obvious mistakes that many speakers make. This has proven to be a successful recipe. If you really know what errors you should avoid, the chances are high that you will be able to greatly improve the effectiveness of your presentations.

### 3.3.1 The attention curve – help the audience to focus on your message

The average participant in a conference is generally willing to listen to you but also easily distracted. You should realize that only a minor part of the audience has come *specifically* to listen to *your* talk. The rest is there for a variety of reasons, to wait for the next speaker, to get a general impression of the field, or whatever.

Figure 3.4 illustrates how the average audience pays attention during a typical presentation of, let's say, 20–30 minutes. Almost everyone listens in the beginning, but halfway, the attention may well have dropped to around 10–20% of what it was at the start. At the end, many people start to listen again, particularly if you announce your conclusions, because they hope to take away at least something from the presentation.

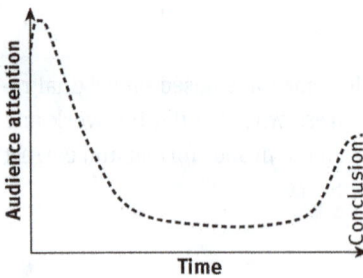

Fig. 3.4: Typical attention the audience pays to an average presentation.

What can you do to catch the audience's attention for the whole duration of your talk? The attention curve immediately gives a few recipes:

- Almost everyone listens in the beginning. This is *the* moment to make clear that you will present work that the audience cannot afford to miss.
- If you want to get your message across, you should state it loud and clear in the beginning and repeat it at the end.
- The best approach, however, is to divide your presentation in several parts, each ending by an intermediate conclusion (see Fig. 3.5). People who got distracted can always easily catch up with you, particularly if you outline the structure of your talk in the beginning.

### 3.3.2 Why are audiences distracted?

There are many reasons why audiences may be distracted. Some reasons may be outside your control, such as inadequate sound systems, poor projectors, too much stray light from the windows, the catering service entering with fresh coffee and tea, uncomfortable seats with chairs standing too close, or inherently noisy conference cen-

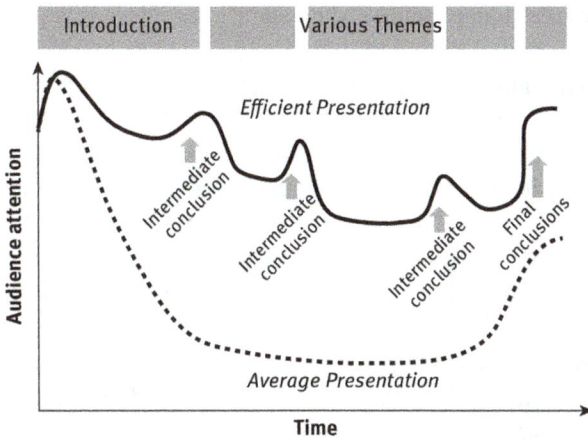

Fig. 3.5: Ideal attention curve of an audience when the speaker divides his/her talk in recognizable parts, each summarized by intermediate conclusions. When people lose their attention for some reason, they can easily catch up with the speaker in one of his/her intermediate summaries. A major advantage of this approach is that every important item is said several times. Repeating the essentials is the key to getting your message across.

ters with sliding walls that partition larger halls so sessions can run in parallel. Bad luck if it happens to you and affects your talk.

What you can do yourself, however, is avoid anything that may encourage the audience to stop listening. Such mistakes fall in two classes: speaker's errors and presentation errors.

We list a couple of the most common mistakes; most are self-explanatory.

### 3.3.2.1 Mistakes you want to avoid

**Audiences love background, but avoid stating the obvious . . .**
You can raise the interest of attendees who are not especially interested in your subject by giving them the impression that they will learn something from your talk. Note that this part of the audience is more interested in general aspects than in details. You certainly need to give them a good introduction into the background of your subject before they can fully appreciate the subtleties of your work. Hence, you should feel free to spend 20–30% of your time on a proper introduction, without, however, resorting to common knowledge (fossil fuels that will run out, $CO_2$ being a threat to mankind, etc.). You don't want to annoy your audience by stating the obvious. However, with sufficient background information that is to the point, they will understand a lot more about your specific results, i.e., that part of the talk you are most proud of.

– Overestimation of the audience: The speaker lives in his/her own little world of research, believing that all the background information needed to appreciate his/her work is common knowledge. Students often assume that the audience knows as much as their supervisor does, which certainly is not the case! Hence, a topic needs to be properly introduced, depending on the specific audience it is presented for.

– The structure of the presentation is unclear, and consequently, the line of reasoning is difficult to follow. Important matters as problem identification, aims, or motivation are insufficiently clear. Sometimes, speakers focus on *what* they have done and *how*, and they forget to tell *why*. See also the next section on how to organize a presentation.

– Visual aids (PowerPoint slides) are inadequate, too busy, confusing, hard to read, too small, etc. Some speakers show too much in a too short time (one slide per minute is not bad as a rule of thumb).

– The speaker uses long, complicated sentences, or unnecessary jargon, abbreviations, or difficult words. Passive sentences ("From this figure, it was deduced that . . ." or "It was therefore concluded that . . .") sound stiff and are more difficult to follow than the active ones ("This figure – point at it – implies that . . ." or "Therefore, we conclude that . . .").

– Even worse is when the speaker reads his/her speech from paper and forgets that written language is usually more formal and complex than language used in everyday conversation. In addition, reading out written text tends to go a lot faster than telling something in a natural way. Sometimes, politicians suffer from this problem, especially when others have written their speeches. In such cases, the audience will for sure experience information overload, and probably, they will give up and stop listening. Of course, we sympathize with the speaker who feels insufficiently confident in English. However, reading out your text from paper is almost always an unsatisfactory solution. After all, nobody in the audience will blame you for a couple of mistakes in the language. Don't forget that English is a foreign language for many in the audience.

– Monotonous sentences, spoken either too fast or too slow, lack of emphasis, unclear pronunciation, slang, and colloquial and fashionable expressions (especially if English is the speaker's native language!) all make it difficult for listeners to stay attentive.

– Some speakers turn their back to the audience and watch the projection screen while they are talking, instead of trying to make visual contact with the audience. Sometimes, the reason is that they use the slides to guide themselves through the talk, or they read literally what is written on them. Know that PowerPoint has a presentation mode where one can see the current slide, the timing, plus some speaker's notes on the computer screen while presenting, which is a better solution than looking at the screen all the time. However, never forget to regularly look at the audience and establish contact. Presenting means communicating!

**Not too fast, please . . .!**
Many speakers have rehearsed their talk so often that in the presentation, they speak too fast. Others simply have so much to cover that the only way to stay within the allotted time is to speed up. Of course, this is not in the interest of the audience, particularly not at an international meeting.

**. . . and try to vary your pace**
As a rule of thumb, speaking at 150 words per minute is all right. However, try to vary your rate. Key ideas, complicated points, or concluding remarks (you may want to use one at the end of every slide you show) are best presented at a slower pace.

### 3.3.2.2  How to organize your presentation – not as a written report!

You should be aware of the fundamental differences between an oral presentation and a written report. In a presentation, the listener, by necessity, will follow the order in which the speaker presents the material. The reader of an article or report can skip parts, go back to the materials section, take a preview at the conclusions when reading the results, etc. Exactly because of this reason, all scientific reports follow the generally adopted structure of *Abstract – Introduction – Experimental Methods – Results – Discussion – Conclusions – References*. However, this structure is hopelessly **unsuitable** for an oral presentation. Nevertheless, many contributed talks at conferences adhere to it, and several universities actually teach it to their students!

Why is this generally accepted report structure unsuitable for oral presentations? Because the listener will have to remember details about, e.g., the experimental methods until the results are presented, and must recall the various results when the speaker deals with the discussion, etc. In other words, details that should be combined (the why, how, what, and what does it mean, of a specific experiment) are treated separately. You are asking a lot from the audience if they need to remember all these facts and figures, until at the end, you finally explain how these bits and pieces fit in a larger picture!

Grouping together what belongs together is the best principle for organizing your talk. Hence, if you discuss materials characterization, you start this section of the presentation with a few introductory remarks of what you want to learn about your catalyst and how the method of choice may help you to provide this information. Then you show some key results and you discuss their meaning. End with a partial conclusion on this section. Next you go to the following item in your presentation, which may be the chemical reactivity of the materials, or so. When you finished this section, you may give an overall conclusion on the state of your catalyst before you go on to speak about catalytic behavior.

In conclusion, feel free to organize your talk in the way you think it will be understood best, and please, do **not** follow the article structure.

### 3.3.3 Ten steps to a successful presentation

The two key issues in preparing a successful talk, summarized in Fig. 3.6, are the following:

- The message: What do I want the audience to know when I am finished?
- The audience: How do I present my talk such that the audience will understand and remember what I want to tell?

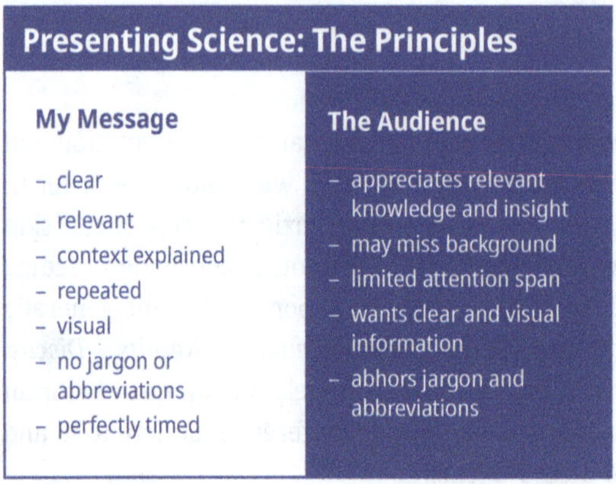

**Presenting Science: The Principles**

| My Message | The Audience |
| --- | --- |
| – clear | – appreciates relevant knowledge and insight |
| – relevant | – may miss background |
| – context explained | – limited attention span |
| – repeated | – wants clear and visual information |
| – visual | – abhors jargon and abbreviations |
| – no jargon or abbreviations | |
| – perfectly timed | |

Fig. 3.6: The principles of communication: your message for a specific audience.

### Step 1: start in time
Once you submitted the abstract to the conference organizers, it is time to start thinking about how you organize the material in a talk when your abstract will have been accepted. Read about the background of your work, read related work, look at your own results regularly, and think about the most relevant conclusions. Try to imagine what type of audience you would have and consider what you would have to include as background information.

### Step 2: the message in a sentence
Try to capture the essential message of your presentation in a single sentence. This is difficult. You will only be able to do this if you really master your subject (which, in fact, is the main requirement for being able to clearly present your work to others). Such a "message in a sentence" is also a good basis for a 30-s elevator pitch, which we will discuss in Chapter 4 (although it is not the same).

### Step 3: select your material and order it
Use the "message in a sentence" under step 2 as the criterion to select which results to include, in what order, what basic information is needed to appreciate these results,

and which experimental details are necessary and which are not. Be very critical, any experiment or result that does not contribute to your main message, however nice it is, should be left out.

Although it may at first sight seem natural to present your results in the chronological order in which you obtained them, this does not have to be the most ideal order for the audience to understand what you have done. Think about where to discuss highlights – at the beginning? Near the end? Maybe you could disperse remarkable features through the entire talk? It is up to you, but apply the order that you believe will appeal most to the audience.

The scientific background of your audience determines how much you should explain about experimental approaches and characterization techniques. Again, be careful *not* to identify your audience with your supervisor; most of your listeners are unlikely to possess much specific knowledge about your subject. By the way, if you explain it well and possibly in an entertaining way, hardly anybody will mind hearing something he/she already knows.

### Step 4: opening and introduction

In the opening, i.e., the first few sentences, you catch the attention, for example, by a scientific question or a catchy or maybe even provocative statement. Perhaps, you could already give the conclusion of your work too. Try to speak slowly, with emphasis, and look at the audience. They may need to get used to your voice and your way of speaking. Of course, it is a good idea to prepare the opening carefully, by writing it out in simple sentences, and to rehearse it several times.

However, before you give your opening sentence, it is good to start with "Thank you Mister Chairman, ladies and gentlemen . . .," followed by a few seconds of silence, in which you look around to establish contact and see if people are paying attention. By doing so, you implicitly get the audience to listen. Starting this way, you also test the sound system, and you ascertain that your important opening lines are going to be heard. In the rest of the introduction, you sketch the background of your research and you introduce your research question, including the reason why it is important. Remember that many people will be interested in a concise summary of the status in your field. Hence, reserve sufficient time (20–30% of the total time) for the general aspects of your work. It is good practice to not only clearly identify the scientific question you address but also give the conclusion of your work, if you wish so. This way, you enable the audience to better follow your reasoning and to anticipate on the outcome of the experiments. In other words, you give them a chance to listen actively. Remember that a scientific presentation is not a detective story that is solved at the conclusion.

### Step 5: conclusions and ending

Conclusions should be properly announced to regain full attention. Present your conclusions in relation to the questions you raised in the introduction. Avoid all irrelevant details. Once you finished the conclusions, you may acknowledge people who helped you (not the coauthors listed in the program) and the funding agencies. Then comes a final opportunity that many speakers miss: end by repeating the message of your talk, for instance: "Ladies and gentlemen, I hope I have convinced you that . . ., etc." This is the take-home message that the audience should remember, hopefully in combination with your name and affiliation.

---

**Don't do this**

An often heard, but poor start is:

*"Good morning, ladies and gentlemen. I am John Talker and **today*** I'd like to tell you something about my PhD project at the Group of Archaic Research at the University of Science City. The title of my talk is . . . . I will start with an Introduction, then explain the experimental techniques, next present the most important results, and finally I hope to draw a few conclusions and I want to acknowledge a few people. So let us start with the Introduction . . ."*

If you open this way, you will find yourself in the company of many others. Nevertheless, this is a totally inefficient way to start a presentation. You are simply telling what everybody has read in the program and was already stated by the chairman. How would you respond if you were in the audience?

*Why say "today" in the opening sentence? Do you give another talk tomorrow? Avoid, it sounds silly.

---

### Step 6: crystal-clear figures and schemes have the highest impact

A picture is worth a thousand words. Well, not necessarily. Figures, especially those generated by spreadsheets, may look neat and tidy, but at the same time, they may be real puzzles (see Figs. 3.2 and 3.7).

A good picture to be used in an oral presentation
- is easy to read (large lettering, good contrast);
- explains itself (clear title, preferably a conclusion underneath);
- contains only relevant information;
- has no jargon or difficult codes that the audience needs to translate; and
- has labels on the data and not in lists in the legend.

Hence, when showing a series of spectra or activity curves, you put an understandable label on each curve (not a, b, and c, which are explained in a separate legend!). Avoid reference to samples in codes such as "Sample AX234/a5," which may be handy in laboratory notebooks but should be absolutely forbidden in presentations (and in articles as well).

Using tables with numbers is, in most cases, not recommended in a presentation. Remember that an audience will generally read everything you show on a slide, and while they read, they pay less attention to what you say. Also avoid too many theoretical formulas and mathematical derivations. Sometimes, you may have to show one, but try to keep it to a minimum. You should realize that our brains remember graphical and pictorial information best. Hence, clear figures, schemes, and diagrams are the preferred means to convey information.

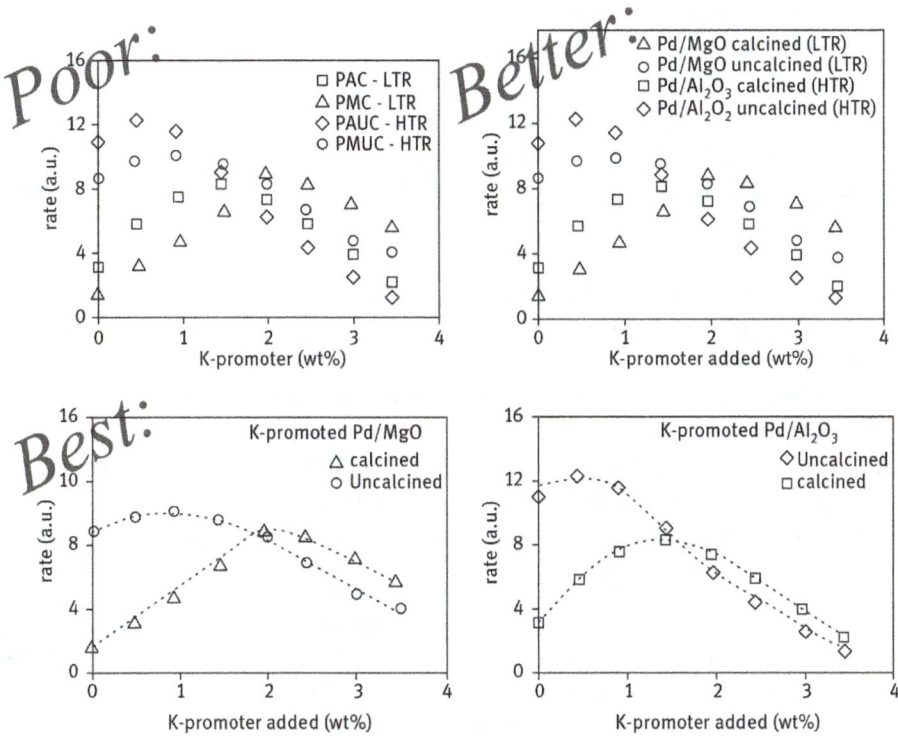

**Fig. 3.7:** Spreadsheets and presentation programs often produce unsatisfactory figures, particularly with respect to labeling. A good figure has labels on the curves and not in a legend. Secret codes and jargon should be avoided as much as possible.

## Step 7: PowerPoint slides

**Tips for effective slides**
- Use large lettering.
- Aim for contrast: Black letters on a white background, bright yellow on black or dark blue, etc.
- Do not use structured backgrounds and do not waste too much useful space on logos, fancy tem plates, etc.
- Use pictures, figures, a title, a short, clear caption, and conclusion.
- Avoid data in tables or inside text.
- If you use text, then avoid complete sentences but use "headline" style as the newspapers do.
- Give each slide a meaningful title and try to include a brief conclusion.

PowerPoint is both a blessing and a curse. If used properly, slides are a marvelous aid to present your work. They also leave you the flexibility to make last-minute changes and to adapt to the situation. Unfortunately, many speakers misuse their slides by doing the following:

- showing too many in a short time,
- overloading slides with data and text,
- using too small fonts,
- showing confusing graphs full of secret codes,
- exaggerating the use of animations,
- forgetting to include a title (aim) on each slide and intermediate conclusions along the way.

In addition, the usual presentation software offers so many inviting opportunities that speakers often use ineffective color combinations and disturbing background patterns.

The best advice on making effective slides is probably to remove all information from slides and figures that is not strictly necessary. Nevertheless, do provide clear understandable labels on curves and spectra in figures, so that they become self-explanatory to the audience. Each slide should contribute to the overall message of your talk, and therefore it must be completely understandable.

## Step 8: think in terms of communication rather than performance on stage

Your presentation will be most effective if you use the same everyday language in which you explain things to a fellow student in the lab. There is absolutely no need to use a more formal language. In fact, formal language is not desirable at all as it is more difficult to understand for the audience. Do not try to impress the audience with fancy words, formal constructions, subject-specific jargon, or unnecessary abbreviations. Think about oral presentations in terms of communication and do not see it as a performance in a theater. The audience will be grateful if they can easily follow what you say, and they will not at all blame you if you make some mistakes in grammar or wording.

## Step 9: timing is absolutely necessary

Now comes the moment of truth: does everything you prepared fit within the available time? There is only one method to find out: take your stopwatch and practice it. This is usually a frustrating experience. First, you may note that the sentences simply do not come. As said before, my solution is to sit down and write the first part out in clear, short sentences. Second, you will probably find that you have too much material. Hence, you should cut down, and we do hope that you will not take out too much of the general introduction. With the attention curves of Figs. 3.1 and 3.2 in mind, it is probably the best to skip a few less important items in the middle of your talk. You should rather not compromise on the introduction and certainly not on the conclusions! Carefully timing your presentation is extremely important. Going overtime is an offense to the audience and to the speakers following you, particularly if there are parallel sessions. Nothing is more embarrassing than that the chairman must stop you before you have been able to present your conclusions!

---

**Don't lose time at the start**

Many speakers, even experienced ones, unnecessarily lose time in the first minutes.

Assuming the chairman introduced you properly, there is no need to restate the title and explain who you are or where you come from. Showing all this information on the first slide is more than sufficient.

Others have noticeable difficulty to get started. Apparently, the intended introductory statements do not come as spontaneously as hoped, maybe because of stage fear.

Note that a good start of the talk is critically important in catching the audience's attention; you don't want to take any risk here. Hence, the best advice to speakers is to meticulously prepare for the first five minutes. Write this part out in short, powerful, crystal-clear sentences and rehearse them several times.

You may do the same for the conclusions section; these are the last sentences the audience hears and hopefully remembers.

---

### Step 10: dealing with stage fear – are you nervous? Hopefully you are!

Only very few of us have been born as a talented speaker. Almost everyone will be nervous before a presentation. For beginners, nervousness may easily lead to lack of confidence, caused by feelings of being inexperienced. However, more experienced speakers know that being nervous is a normal part of the game, and nothing to be afraid of.

First-time speakers often interpret nervousness as a sign that they are apparently incapable of delivering a good presentation. This is not true. All the symptoms that accompany nervousness, such as frequent swallowing, trembling, transpiration, etc. are signs that your body is getting ready for something important. Athletes, stage performers, musicians, and experienced speakers have learned to recognize these symptoms and to appreciate them. They start to worry when these symptoms stay away!

Experience will come in time, by practicing and by evaluating your presentations and those of others. No one in the audience will blame you for being a beginner. However, if you take care to avoid several typical mistakes that beginners (and even experienced speakers) make as discussed above, you will make a very good start with your career as a presenter. If you know and understand the basic principles and you know how to apply these, you are likely to give a talk that is significantly better than the average presentation at international meetings. Hence, lack of experience is not important, provided you prepare your presentation well and you do your best to avoid the obvious mistakes.

In conclusion, giving a successful talk at a conference, a workshop, an instruction course, or just in a work meeting of your own research group, all depends on how well you are aware of the two main principles:

- *What is the message I want to convey?*
- *How does the audience understand and remember my message best?*

Awareness of how audiences listen and memorize is the key to a presentation that will be appreciated by many. Learning to apply these two principles is the best recipe for giving effective talks. Feel free to develop your own style, but avoid the usual mistakes so many speakers make.

## 3.4 Posters: ideal enablers of discussion

A successful poster conveys a clear message through high-impact visual information and a minimum of text. Posters are very important vehicles for presenting and discussing work at conferences. Poster sessions provide a wonderful forum to meet colleagues and discuss scientific work on a person-to-person basis. Unfortunately, a fairly large number of posters do not succeed in drawing significant attention. Here, we list some of the most frequent mistakes that presenters make and we make some recommendations for composing efficient posters.

### 3.4.1 When is a poster successful?

A poster is successful if it conveys a clear message to the visitor and generates discussion and valuable feedback to the presenter. To achieve these goals, the poster needs to be crystal clear about the objectives, the approach, the main results, and the major conclusions of the work, and all this preferably within the proper perspective of existing knowledge on the particular subject.

### 3.4.2 Frequently made mistakes

Too many posters do not succeed in getting their message across. Here are some of the main errors presenters make:
- **Too much text.** Roughly more than half of all posters have way too much text. Posters containing 1,000–2,000 words (i.e., almost a complete publication!) are no exception! Whom of your visitors will take the time to read all this?
- **Unclear structure.** If key elements such as objectives, approach, conclusions, or perspectives are missing, everyone who is not an insider on your subject will not understand why the poster is relevant (and why he/she should spend time on it).
- **Inappropriate structure.** Many people blindly apply the standard structure of a written report (introduction – experimental – results – discussion – conclusion), thereby using their poster as a sort of miniature article, which almost automatically leads to a lot of text. There is no standard structure for a poster, just design a layout that works.

– **Poor figures.** Some figures may be real puzzles, with incomprehensible legends, secret codes, small lettering, cryptic captions, etc. As said earlier, spreadsheet and data programs generally do not produce "reader-friendly" graphics.
– **Too much information.** Many presenters overload their posters and greatly overestimate the time that the average visitor is willing to spend on the poster.
– **No presenter present.** This is a missed chance for getting feedback. Another frequent mistake is that presenters take a passive attitude and make no effort to initiate discussion. A presenter who is busy with his/her smartphone does not invite interaction.

### 3.4.3  Ten steps to a successful poster presentation

1) **The message of your poster.** Try to formulate the essence of what you want to present in a single sentence, the message in a sentence. Examples of such sentences are as follows:
    – *I want to convince the audience that my new synthesis method produces a new material of outstanding purity and definition we seek ways for further improvement.*
    – *Analyzing reaction X with our advanced and improved micro-kinetic model is the key for exploring commercialization of this important process.*
    – *The novel ABC technique yields unexpected but beautiful structures of molecules – but we do not understand why these can be stable!*
    The last mini-message might generate some interesting feedback on your poster. Use this sentence as a guide for selecting the data you need to include and the conclusions you want to stress. You probably will not actually print this sentence in the poster, but it helps you to make up your mind what your poster is about.
2) **Introduction.** Write a few brief sentences of introduction (maybe in newspaper header style?) to identify the problem you address, why it is important, what is known about it, the objectives of your work, and your approach to investigate it. Use short sentences and keep this section as concise as possible. Consider if a bulleted list or a graphic might be used.
3) **Results.** Select only the most pertinent results that support your message and remove everything that is not absolutely necessary. Think about attractive ways to present the data in figures. Try to avoid tables as much as possible. Figures and captions should be easy to read – remove all abbreviations and secret codes and use only understandable labels. If possible, add a brief conclusion to every figure, so that there is no misunderstanding what it means (according to you).
4) **Conclusion.** Write the conclusions in short, clear statements, and maybe as a list. Finish with an assessment of what you have achieved in relation to your objectives and, perhaps, what your future plans are.

5) **Attention getters.** What are you going to use to draw people's attention? An attractive title serves as such to some extent but is not enough. Select one of your most important results, a photo or a scheme explaining the scientific background, a model or the main conclusion, or whatever you consider as the highlight of your presentation and give it a prominent place on your poster. This is what the audience will see first. It should raise their interest and stimulate them to read your poster.

6) **Layout.** Arrange all the parts of the poster around your attention getter. Add headers if necessary to clarify the structure of your poster, and add everything else that is needed, such as literature and acknowledgments. Ensure that author name(s) and affiliation(s) are on the poster and that you acknowledge who needs to be mentioned.

7) **Review, revise, optimize.** Ask your coauthors and/or colleagues to comment on a draft version of your poster. Assess very critically if the poster indeed conveys the message you want.

8) **Presentation.** Have a "message in 30 s" ready for people who walk by and read the title of your poster but have not decided yet if they want to go through the entire story. Prepare a somewhat longer version (1–2 minutes) for those who are interested in the subject. The sooner you get into a question-and-answer mode with the visitor, the more you are likely to benefit from the interaction.

9) **Be proactive.** Prepare yourself for the poster event by studying the list of participants. Are there key scientists whom you would like to discuss the poster with? There is nothing wrong with approaching such people before and asking them kindly to visit the poster because you would be very keen on discussing a certain issue with them. You may, for example, give your business card, with the poster number written on it.

10) **Follow-up.** Consider giving selected visitors an A4-sized reprint of the poster if they show genuine interest. If you already published on the subject, it may be useful to send a PDF in an email with a short "thank you for your interest at the recent conference." (See also the section on contact and relation management in Chapter 4.)

In conclusion, a successful poster conveys a clear message through high-impact visual information and a minimum of text. A good poster enables the visitor to grasp the message in a short time, e.g., less than a minute. If the visitor finds the subject of interest, he/she will stay to learn about the details and discuss the work with the presenter. If you fail to get the reader's attention in a short time, he/she is likely to go on to the next poster, unless he/she really wants to know about your work. Search the internet for examples (good or bad, you learn in any case).

## 3.5 Conferences

Conferences, symposia, seminars, workshops, or simply scientific meetings, all of them are ideal events for sharing, exchanging, and getting new ideas for your research, interacting with others, and becoming an active member of your scientific community.

There seem to be ever more conferences every year. "Too many meetings" is the complaint of many academics. Nevertheless, many eagerly engage in organizing new ones or extending that one-time successful meeting into a periodic event. For you, having the organization of a successful symposium, workshop, or summer school as an achievement on your CV will certainly be a point in your favor when it comes to promotion or qualifying for a position elsewhere. The purpose of this brief section is to make you aware of the different types of meetings and get you thinking if you could play a role in organizing one. We will not go through the entire scenario of planning a conference but refer to some of the excellent resources, in print (Epple 1997) or available on the internet (Key; Work Group for Community Health and Development; School of Advanced Study).

### 3.5.1 Types of meetings

Conference and congress are the most general terms for meetings with a program of oral and poster presentations, discussions, and sometimes exhibitions and demonstrations, in a program of plenary and parallel sessions, extending over a few days. These events are often organized by or under the auspices of a scientific organization and held on a periodic basis. A local organizing committee is in charge, but sometimes it follows a general protocol furnished by the parent organization. Examples are the conference series under the European Physical Society, the semi-annual National Meetings of the American Chemical Society, and many others. Such meetings are often visited by several hundreds to thousands of participants and are, in practice, a collection of smaller topical symposia. The large-scale ones offer many attractive features, such as a sizeable exhibition with the latest instrumentation, books, and journals, while the constituent symposia serve as the focal points where subcommunities meet and interact.

**Beware of predatory conferences**! These have only one purpose: to make money. Such events, often organized in attractive locations, charge high participation fees but try to seduce you into participation with flattering invitations for an invited lecture, a keynote address or a regular contribution, offering a reduced fee If you accept. Presentations are generally accepted without peer review, resulting in an incoherent program of variable quality, with limited scope for useful discussions. The chances to meet interesting colleagues are less – experienced scientists will not go there. We strongly recommend you focus on meetings organized by the "learned societies" and reputed scientists. Budgets for visiting conferences are often slim, don't waste them on expensive trips to worthless meetings in tourist hot spots.

Although these large meetings can be overwhelming, they do offer the opportunity to sample what is going on in the field, in addition to the specialized symposia of your own interest. The disadvantage is that it can often be quite difficult to find and meet people and that most participants tend to have busy schedules, trying to follow talks scheduled in different halls of a large convention center (or sometimes even in different hotels). Another problem could be that the program features too many symposia, with some ending up at unfavorable time slots (e.g., Sunday morning, Friday afternoon), or receive less attendance than expected. As a rule of thumb, such large conventions have peak attendance during the first two or three days, after which attendance tends to fall off.

**Invitations to speak: accept or decline?**

Many scientists receive several invitations per year to speak at scientific meetings, varying from large international congresses to national meetings in their field of research, or educational events such as summer schools. Conditions vary from all expenses including travel and accommodation paid, to reduced registration fees, or even no financial advantages at all. Use some of the following criteria to decide:

- Topic(s) of the conference.
- Is the organizer a respected scientist on the topic, and is there a respectable advisory board with known scientists?
- Who are the other invited speakers?
- What is the intended audience and is it important that they hear about your work?
- Are you obliged to contribute an article to the proceedings, and if so how will these be published? (Many colleagues decline if an article is requested.)
- What are the financial conditions (registration, travel, accommodation)?
- Does the event fit in your agenda, or does it interrupt other activities too much?
- Can you perhaps send a young coworker instead to gain experience?

In any case, carefully consider if accepting a seemingly flattering invitation is a good idea.

Workshops are much smaller than conferences and can have anywhere between 10 and 100 participants. Workshops are usually devoted to one specialized topic, e.g., a specific research methodology, a specific technique, or a recent discovery in the field. The typical format of a one-day workshop is as follows: one or more expert lectures of 30–45 min each, several slots of 10–15-minute presentations, ample time for discussion, both plenary and in breaks, room for posters, and an experienced participant who summarizes the discussions at the end of the workshop. Such meetings are relatively easy to organize inside your own university. The costs are usually limited to catering and perhaps some reimbursement of a star speaker, and the fee for participants will be relatively modest. (*Note*: Preregistration and payment of a fee are usually a good idea, as participation free of charge may result in people not showing up at the last minute.)

**Summer schools and discussion meetings** are our favorite type of meetings, as they tend to be more relaxed, with better opportunities for in-depth discussions in a pleasant setting. Often aimed at providing a high-level overview of what is hot in an area, such meetings offer a program with experts who are given ample time to present their recent work, usually in the context of state of a field. Participants can often bring a poster about their work, to present in an evening session (discussion not seldom being facilitated by ample beer and wine). Gordon Research Conferences are a famous series of such meetings in the United States, while Europe has a rich tradition in organizing summer schools aimed at high-level education of PhD students. Often, such meetings take place in the "quiet" rural settings of monasteries, schools in vacation time, or vacation centers outside the tourist season.

A rather unique type of meetings in this category is the Faraday Discussion, organized regularly by the Royal Society of Chemistry of the UK. Here, speakers contribute a manuscript, which is peer reviewed. When admitted to the conference, the papers are distributed to all registered participants well before the conference so that they can read it. During the conference, each contributor summarizes the manuscript in five minutes to fresh up the memories of the audience, after which a lengthy discussion of about half an hour follows. Questions and answers are recorded and published along with the article after the conference. In fact, the collection of discussions on a series of related papers is published as a separate article, with coauthorship of all who submitted a question or a comment. This unique type of meeting gives rise to thorough discussions and exchanges of views, while a large part of the audience is actively involved.

**Minor issues can have major consequences and spoil the meeting**
In conferences, little details of seemingly minor importance that spontaneously go wrong or have been planned improperly can turn into the talk of the day and negatively affect the atmosphere. We remember a European conference visited by some 1000 participants, where daily lunch was arranged in the form of two sandwiches, snugly packed with an apple, a drink, and a napkin in a paper box – for some an efficient time saver, but a culture shock for many international participants longing for the opportunity to socialize in a relaxed setting over a decent three-course meal with wines. Many years later, seniors still talk about the "lunch box scandal," but only a few remember the excellent program.

## 3.5.2 Opportunities – engage!

Scientific conferences are a marvelous source of information on what is going on in the community, for getting new ideas for your research, meeting your peers, presenting your own work, and getting feedback. But it requires that you, and the coworkers and students you bring along, are active. All too often, we see that conferences are regarded as a sort of school outing, where during coffee breaks, lunches, and other social events, students hang out mostly with their own group – people they already see every day! This is a missed opportunity, of course. Educate them and give them a good example.

Organizing a symposium at a conference, a workshop at your institute, or a summer school is a great opportunity to get experience, while you also provide valuable service to the scientific community. Provided you do it well, you raise your visibility and you will certainly make worthwhile new contacts. Many organizations offer the opportunity to propose topics for symposia in their periodic conferences. Maybe you can team up with a somewhat more experienced colleague, make a proposal for a symposium, and hopefully have the chance to organize it – a very valuable experience, and if you do it well, you will be remembered.

Some journals see such specialist symposia as a good occasion to publish a special issue on the topic of the meeting with invited contributions. Why not propose a thematic issue to such as journal as well? Serving as guest editor is another great opportunity to learn and at the same time raise your profile in the community.

### 3.5.3 Finally, some sources of irritation

Once again, it is not our intention to give a crash course into organizing meetings. Nevertheless, we would like to finish this section by sharing some frustrations from our experience acquired over 45 years in at least 300 scientific meetings of all sorts. Table 3.5 shows a list of personal irritations and things that sometimes go wrong in a conference.

Tab. 3.5: A few issues at conferences from personal experience.

| What sometimes goes wrong . . . | . . . and how it should be |
|---|---|
| **Program** | |
| "Routine" programs reminiscent of last year's conference, same faces on the podium as everywhere else, especially among the invited speakers. | Balanced program of a few high-profile speakers for perspective and many (young) people presenting new work. |
| Overloaded with oral contributions, leaving too little time for breaks and poster sessions. | Breaks every 90 minutes at least, adequate attention for poster presentations and discussions; the main value of a conference is networking, not consuming knowledge. |
| Long opening session with irrelevant speeches; sometimes the local guest of honor speaks no English. | Limit opening and closing formalities to the absolute minimum, rather create more time for contributed talks or posters. |
| Parallel symposia not synchronized; people switching sessions miss important parts. | Strict timing is crucial, especially with parallel sessions. |

Tab. 3.5 (continued)

| What sometimes goes wrong . . . | . . . and how it should be |
| --- | --- |
| **Poster sessions and exhibitions** | |
| Too many posters for too short a time in too little space, sometimes without sufficient light; posters running during lunch and coffee breaks only; presenters not at their posters. | Take poster sessions seriously; they offer valuable opportunities for discussion and exchange of ideas, as well as networking. Keep posters on display for the entire meeting and consider making the posters/exhibition venue the social center of the meeting (a bar makes a huge difference). |
| Insufficient space or unattractive site for exhibitors; no time in the program for giving proper attention to exhibitors. | Exhibitors often contribute significantly to the budget and help to limit the registration fees; make the event worthwhile for the exhibitors and for the audience. |
| **Organization** | |
| Name badges too small, names cannot be read. | |
| Long queues for registration or for coffee/tea and toilets in the breaks. | Prepare well for logistical details that affect the atmosphere of the meeting; don't surrender too easily to catering services and venue owners. |
| **Venue** | |
| Lecture halls filled up with uncomfortable chairs, sometimes placed too close to each other. | Ensure comfortable seats with enough room; a lay out with table space for every attendant is ideal (though not always possible, of course). |
| Noise, sometimes caused by loud conversation outside the halls, by catering services, or by inadequate flexible dividers. | Maybe a member from the organizing committee can be assigned to limit such disturbances to the minimum. |
| Workshops in venues with U-shaped table arrangement, causing stiff necks by having to turn heads toward the front (with views blocked by neighbors). | Audience should have unobstructed views at the podium. |
| **Equipment for presentation and sound system** | |
| Inadequate sound systems, lack of microphones. | |
| Failing presentation equipment, weak laser pointers, or lack of spare batteries. | |
| **Audience** | |
| Mobile phones going off – or even being answered – in the audience; people tweeting, emailing, texting, typing loud; people talking during presentations. | It all seems to indicate that people do not come in the first place to listen and acquire new knowledge, but that meeting face-to-face with others is the most important function of a conference. |

Tab. 3.5 (continued)

| What sometimes goes wrong . . . | . . . and how it should be |
| --- | --- |
| **Chair persons** | |
| Sometimes unprepared, stumble over titles and names, or forgot to make acquaintance with presenters before the session. | Prepare, read abstracts and titles, and check pronunciation of names while meeting the speakers before the session. |
| Forgetting to enforce timing, at the cost of question time or even disarranging the entire program in this way. | Traffic light systems switching to yellow a few minutes before the speaking time is over often work well, red means absolute stop. |
| Allowing questions from seniors only or mostly from the first row. | Try to engage the audience in the discussion; reserve the first questions for students. |

# References

Boxman, R., Boxman, E.: Communicating Science, a Practical Guide for Engineers and Physical Scientists; World Scientific Publishing Co., Singapore, 2017.

Dodd, J.S.: The ACS Style Guide; American Chemical Society, Washington, DC, 1997.

Epple, A.: Organizing Scientific Meetings; Cambridge University Press, Cambridge, UK, 1997.

Garcia, H.E.: The Power of Communication: Skills to Build Trust, Inspire Loyalty, and Lead Effectively; Pearson Education, Upper Saddle River, NJ, 2012.

Garfield, E.: Citation analysis as a tool in journal evaluation. Science; 1972; 178; 471–9.

Glasman-Deal, H.: Science Research Writing for Non-Native Speakers of English; Imperial College Press, London, 2010.

Hirsch, J.E.: An index to quantify an individual's scientific research output. PNAS; 2005; 102; 16569–16572.

Key, C.: The Keynote Guide to Planning a Successful Conference; Keynote Networks Ltd. https://spie.org/documents/students/conference_guide.pdf.

Moed, H.F., van Leeuwen, T.N.: Impact factors can mislead. Nature; 1996; 381; 186.

School of Advanced Study, University of London: Organizing a Conference, Postgraduate Online Research Training. http://port.modernlanguages.sas.ac.uk.

Scott, S.L., Jones, C.W.: Superlative scientific writing. ACS Catal; 2017; 7; 2218–2219.

Smith, R.: Introductions in How to Write a Paper. In: Hall, G.M. ed. BMJ Publishing, London, 1994; pp: 6–15.

Work Group for Community Health and Development at the University of Kansas: Organizing a Conference, The Community Tool Box. http://ctb.ku.edu.

Strunk, W., White, E.B.: The Elements of Style. Fiftieth Anniversary Edition; Pearson Longman, New York, 2009.

Waltman, L., Van Eck, N.J.: Field-normalized citation impact indicators and the choice of an appropriate counting method. J Informetr; 2015; 9(4); 872–894.

# 4 Management skills for researchers

## 4.1 Introduction: the scientist as manager

As we described in Chapters 1 and 2, being a successful scientist requires activities on several fronts. In his Guest Column (page 97), Graham Hutchings presents his three most important enablers for high-quality research: "great ideas, money, and time." Running a large laboratory that covers several research topics obviously needs a smooth organization to ensure that each subject receives the attention it deserves and the resources it needs to become a success. But the same is true for the much smaller research activity of a young assistant professor or a junior research scientist. Investing in skills that help you organize your work efficiently and effectively is essential, and it is wise to learn at least some of these at the beginning of your career. We therefore devote this chapter to management skills that we believe are particularly helpful for a researcher. You probably don't have to become an expert in the finer details of all these skills, but knowing the principles so that you can regularly think about how these can help you in your work is what we aim for.

## 4.2 Fundraising: who will pay for your research?

Research is expensive. You need a laboratory with a specific infrastructure. A chemist will need a well-ventilated lab with fume hoods, workbenches with electricity, certain gases, flowing hot and cold water, lots of glassware, analytical instruments, safety installations, storage cabinets for chemicals and gas bottles, computers and data lines, etc., while an engineer likely wants a laboratory where he/she can design and build models, demo systems, and pilot plants, perhaps supported by a small workshop. A theoretician needs computers, space to put them, and ultrafast data connections to supercomputers or personal computer clusters, and all need office space and desks for people to work on. You most likely also benefit from central services, which must be paid for. If you are lucky, your university provides much of this, but many times, you will have to arrange the funds to support students and postdocs, to buy specific instrumentation and consumables, and to travel to conferences.

### 4.2.1 How are universities and research groups funded?

It is useful to have some insight into how universities are financed and how your own research group fits within this structure. Although variations across different places in the world and even across institutions within one country may be large, the scheme in Fig. 4.1 probably captures the essence of how a university is funded.

https://doi.org/10.1515/9783111325644-004

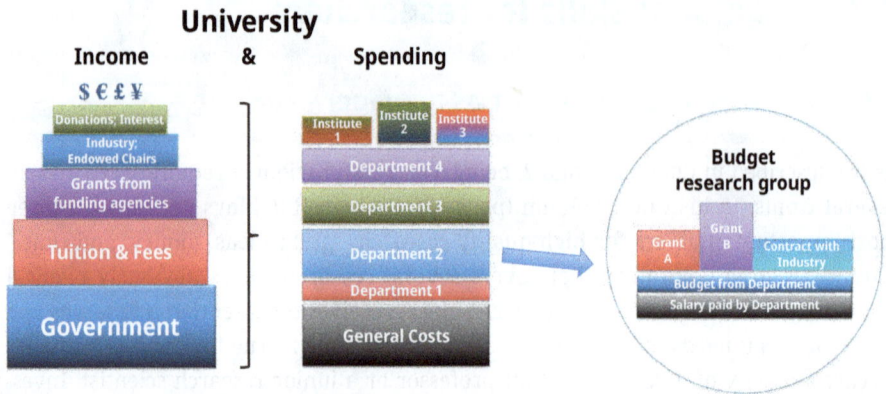

**University**

Income & Spending

$ € £ ¥

Fig. 4.1: Funding structure of a university (please note that variations between different places in the world or even within one country may be large).

Most academic institutions receive an annual budget from the local or national government. On top of this, they charge tuition fees, although there are also countries where education is free and covered by the government. Then there may be income from funding organizations, industry, donations from alumni or philanthropic institutions, legacies, and interest from university capital and dividends from shares. To give an order of magnitude estimate, a typical university in Western and Northern Europe may have an annual budget between a half and one billion euro, while the wealthiest universities in the world might well spend several billion dollars per year.

**Who is the boss of the university?** That depends on the country and its academic traditions. In the USA, the highest position at a university is that of president, sometimes called chancellor, who serves as chief executive officer, while the provost fulfills the role of chief academic officer. In the UK and countries that follow the British Commonwealth tradition, the vice-chancellor is the acting head of the university. Chancellor is an honorary title, often reserved for a member of the royal family or the governor. In the author's country – The Netherlands – the highest position is shared between the chairman of the executive board and the rector magnificus, while some universities combine these roles in one person. In many European and Latin American countries, the rector is the academic head of the university and the highest authority in accordance with the constitution and statutes of the institution.

At the next level are the departments (USA) or the faculties (Europe), led by a head of department, or a dean. In many countries, the word "faculty" refers to the teaching staff with a tenured position. To add to the complexity, universities may group departments into a larger faculty, e.g., the departments of chemistry, physics, and biology, each with their own head in a faculty of natural sciences under a dean. Confusing? Indeed, variations are large. Make sure you understand the organization of your institution and that of any university where you apply for a position.

Each year, the university's executive management (see the text box on government and titles) determines how the money is divided over central costs (administration, buildings, central services, etc.) and the departments and institutes. In a department, the dean or head of department usually decides how the budget is distributed. Maybe he/she uses transparent schemes to allocate money, or they do it with their own wisdom and discretion. Again, it is essential to realize that finance models can vary widely, which will impact the budget of your research group and your freedom to spend it. We give a few examples of how a department can be funded for reference and recommend that you find out how it works in your environment.

Example 1: At mid-sized university A, department X receives its income from the central budget provided by the university, plus the 15% overhead it charges on all external funds (e.g., grants and industrial contracts) brought in by the staff. In this example, the department pays the salaries of all permanent staff (tenured academics, technical, and administrative support). All full professors are in the fortunate position to receive two university-paid PhD students. The central budget covers the costs of teaching, administration, maintenance and use of the building and all utilities, the workshop, the library with journals and books, and the cost of some large, shared research infrastructure (such as an electron microscope or a computer cluster). If research groups receive external grants, they pay 15–25% overhead to the department; the rest is used for the stipend or salary costs of their nontenured coworkers and the running budget for consumables, travel, and perhaps some investments. Newly appointed academic staff usually negotiates with the dean that they receive start-up funds for equipment, extra PhD students, or a postdoc. Note that all this represents just an example of how finances may work.

In a scheme like this, which is generally common in Europe, the tenured professors have job security and freedom to determine their own research subjects, as long as they teach well and manage to get research done. However, to run a substantial research program, they will have to successfully bring in funds from agencies such as the National Science Foundation or from bilateral contracts with industries.

Example 2: Department B at another university operates in a way that professors are forced to be real entrepreneurs. The central budget pays for the buildings, central administration services, and teaching. Still, the rest of the budget is distributed among the academic staff, according to a distribution scheme that scales with their accomplishments in the previous years (e.g., number of diplomas, publications, patents, and credit points for teaching) and the generation of funds for their respective groups. The professors can spend the budget on salaries (including their own), equipment, consumables, and travel. However, they must acquire grants and contracts to work and even secure their own jobs in the following years.

The positive effect of this system is that success is rewarded by a more significant share from the central budget, which works as an incentive for success. The downside is that tenure for the professors is no longer a guarantee for job security and that the capacity to bring in funds is more important than academic excellence. Risky situations may arise if too many groups become too successful, and the departmental funds for each group decrease, or if the department becomes overly optimistic in prosperous times and appoints more academic staff than the system can sustain when the funding climate becomes less favorable. In such a case, the department may find itself forced to reorganize and lay off people. We know some examples where funding mechanisms of this type have led to situations where the university is not much more than a "hotel," which offers space and some minimal support, while professors are supposed to generate income needed for all their activities, including teaching.

As mentioned, the two examples represent extreme cases, and several other funding schemes exist. It may take some time before young starting professors understand how it works in their environment We strongly recommend figuring this out as early as possible, for example, during your job interview.

It is pretty normal that candidates who are offered a position negotiate with the university to obtain start-up funding, before they sign the contract. If the university is eager to have you, they are probably willing to help you with some investments, perhaps with some PhD students, and one or more postdocs. Although the space for negotiation tends to be limited these days, you should always bring this up during your application process (and not after signing a contract).

### 4.2.2  How is your research group funded?

The next question is how much base funding the university or, rather, your department will give you each year. Will they just pay your salary (some universities only pay you during the teaching term, i.e., for 9 months), and can you use the laboratory space for free, or are you supposed to share in the costs for space, utilities, and overhead? Can you make use of central equipment or facilities of other groups, and do you have to pay for this? Do you receive a basic budget for consumables, or will you have to find funding for all these costs? Who pays the stipends or salaries of PhD students? Conditions may vary widely over the world.

Inevitably, you will want to go after external funding at some stage in your career, because you need the money for research and you may want to show it on your curriculum vitae as well.

### 4.2.3  Writing an effective grant proposal

We assume you have a viable plan for an excellent research project that fits your profile. This may sound trivial, but too many scientists do the opposite: they look for funding opportunities and then start thinking about what might be a good proposal. Consider that writing a convincing grant proposal takes a lot of time. Do you want to invest this time, or is it better to spend the time writing up your research in articles first to give you more credibility and a stronger position to apply next year?

Orient yourself on the types of funding sources available. Special grants for young researchers may be smaller but are a good stepping stone for larger projects in the future. Some countries have several sources where you can apply. For example, National Science Foundations tend to have funding programs for fundamental research, while there may be separate funds for applied projects together with industry. Carefully read the instructions, and make sure you understand the goals of the funding program you are applying to. Proposals aiming at testing a clearly formulated hypothesis tend to do better in fundamental science programs, while explorative journeys into the unknown but with big rewards if the idea works are better suited for innovative programs. Discussing your idea for a proposal with a program officer is a good idea. These people are generally keen on funding interesting ideas of promising newcomers, and they are usually very open to discuss their programs with you.

Networking and belonging to the community where you plan to research is vital (see also Section 4.6). It is a must to know the scientists in the field where you apply, and they should know you, too. So, be present and speak at their national meetings, or visit some of these groups, and make sure that your peers and the leaders in your area know about your work and your ideas. There may be valuable opportunities for collaboration. Anyway, some of the more experienced community members are likely to be the reviewers of your proposal, so it is essential for you that they have a good impression of your potential. And, of course, it may be a good idea to cite their work if applicable, both in papers and grant applications.

Write the proposal around a clear idea, and keep this as the focus. The "why" of your idea is as important as the "what" and "how." Everything we wrote about the importance of a clear message in presenting science (Chapter 3) is as important here as in a publication, a poster, or a talk. Avoid long technical diversions, and if these are absolutely needed, consider putting them in an appendix or a text box, so that they do not interrupt the main story. This way, reviewers will notice that you know your subject, but they do not have to read it all if they do not want it. Be sure to cite pertinent papers and note the gaps in understanding that you address.

Make sure the proposal is clear for scientists in the broader discipline, but do not assume that they are necessarily experts in your specialization. So, avoid typical jargon, abbreviations, figures, and tables for the insider, and explain also how your expertise fits within the broader discipline and indicate what the state of the art in your field is. Well-laid-out schemes, figures with clear captions, and other attractive elements to break long pages of text may let your proposal stand out among the competition.

**15 min per proposal . . .**
As a reviewer of a series of project proposals, I (JWN) typically spend 10–15 min to read through a proposal. If I do not understand what is proposed, I give up and give a negative recommendation. If the aims are clear, I am happy to invest more time and provide detailed comments for the program jury to base a fair decision on and for the submitter to clarify and improve the proposal, if necessary.

Having the essence of the proposal in terms of key ideas, criteria for success, and requested resources outlined in the first page(s) will help reviewers a lot in their evaluation. Just keep in mind that reviewers are busy people. If the review is organized in panel sessions, the members may have to work themselves through 10–30 of these proposals in a day. Concise and crystal-clear applications will undoubtedly be appreciated, and complicated puzzles will likely be put aside.

Potential risks and pitfalls, or controversies in the field, should be mentioned and assessed, and having an alternative approach if the original one fails may help to give the reviewers more confidence in you.

Avoid asking for unreasonable items. For example, requesting a budget for a one-day-per-week technician or workshop hours may be an excellent idea, but don't overdo it. Justify why you need such extra resources.

Having an experienced professor as mentor, who is willing to listen to your ideas and read your draft proposal in an early stage, can be very valuable. Mentors do not have to come from your own department. With some distance from your subject, they may very well comment on how effectively you explain the essence of your ideas. Feedback is vitally important for you.

Once the proposal is ready, let it rest for a few days (if deadlines allow). When you have reread it, decide if it has a chance to score high. Please don't gamble with a mediocre or incomplete proposal; it may hurt your reputation at the funding agency and with the reviewers, who will probably know you!

Depending on the procedures, you may get the chance to reply to the reviewers and make minor adjustments before the grant organization makes its final decisions. This is a chance to correct misunderstandings or wrongly interpreted parts or provide additional explanations. Our experience is that reviewers who know the field are generally fair and constructive. Still, occasionally, we have also received some nasty reports of people who did not believe in our ideas (or in us). Get angry as much as you like, but when you reply, you are courteous and to the point, and you explain concisely but clearly why the reviewer is wrong. Keep it short, because long replies to referee comments don't go well with juries. If there is too much to correct, there was something wrong, they will probably argue. If you sense unreasonable hostility or evident lack of expertise in a report, then why not call the program officer and ask advice on dealing with the unjustified criticism?

Unfortunately, the good old days when the authors entered the academic world, and more than half of the proposals were granted, are over. Typical success rates have dropped tremendously, sometimes below 5%. The European Union often works with a grade system, where your proposal has to score at least 80% to be eligible for funding. If you end above the threshold (always a happy moment to hear), it does not guarantee that you will get the proposal granted, unfortunately. Nevertheless, you should be encouraged by the outcome. Resubmission of an application that was close to success should always be considered seriously. Maybe you can discuss the outcome with the program officer and do your utmost to improve the proposal for the next round.

## 4.3 Time management – creating quality time to think

Time management is about being effective and making the best of the available time. We always seem to be short of having enough time to think, be creative, and do important things, whatever these are. It is an intriguing dilemma that everyone of us disposes over all the time, 24 h per day, 7 days per week, that is available; nevertheless, we always seem to have too little of it. There are many books on time management that provide simple or quite complicated schemes to help plan and keep track of activities (Hindle 1998). Here we limit ourselves to a few basic and rather practical principles that have helped the authors to manage their time.

### 4.3.1 Priorities: urgent is not the same as important

The first point to realize is that looking at the many tasks and activities we are supposed to do, some are important while others are urgent (see Fig. 4.2). However, importance and urgency are different. Cleaning the laboratory may be urgently needed, because the dean or director will come around with important visitors tomorrow. However, finishing an important grant application before the deadline expires may have much higher priority than meeting the dean's guests; in other words, completing the grant application is both important and urgent. Another example: you need to prepare well ahead of time for a prestigious keynote address 3 months from now. This is very important, but not immediately urgent as you might also choose to start preparations next weekend.

Hence, our first recommendation is to keep a list of tasks and activities that includes priority, urgency, and deadline. Many email/calendar systems enable you to make such a list, along with timely reminders, and indications of how far you already progressed. Ensure that each task's purpose is clear. Make it a habit to start the day by looking at the list, selecting activities to carry out, and by updating the list at the end of the day.

Fig. 4.2: Time management is about doing the right things at the right time. The scheme is also known as the Eisenhower model of time management.

### 4.3.2 Personal quality time zones during the day

When one of the authors (JWN) became an associate professor in 1989 and had several PhD students, he also worked on a textbook. The other author (JKF) made him aware that certain activities are best done during certain parts of the day and that it is perfectly fine to declare such time zones off limits for others. In this way, JWN found out that for him, the early morning hours were best suited for creative activity, i.e., writing, designing figures, studying literature, and thinking deeply about content. Hence, as far as possible, the time until the coffee break at 10:30 am was "official quality time," and the office door remained closed. Students and colleagues knew the pattern and generally did not disturb before coffee. A similarly productive time zone arose at home in the evening, after dinner, when the children were in bed. Conversely, the afternoon was better suited for meetings or typical managerial duties and administrative work.

Discovering your patterns of effectiveness and time zones for specific activities and enforcing these can be a great way to make more of your time.

### 4.3.3 The Pareto principle: the 80/20 rule

It is a comforting and reassuring idea that output is not necessarily linearly related to the effort spent. Applying Vilfredo Pareto's 80/20 rule to time management suggests that 80% of your beneficial production results from only 20% of your time (see Fig. 4.3). There are many ways to look at this, but our experience tells us there is a lot of truth in Pareto's principle. Many researchers will probably agree that from the time spent on a specific project, including all the efforts on building up equipment and experiments, learning how to do things, and trying out ideas, the definitive results that ended up in the final report or scientific publication accounted for only a fraction of the time. Of course, it is impossible to have the top of the iceberg without the underwater part (although policymakers sometimes seem to forget this). In hindsight, one could always have reached the goal faster than it happened.

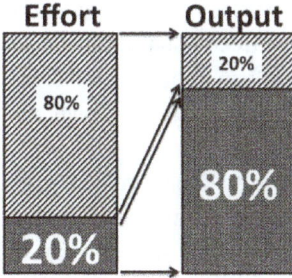

Fig. 4.3: According to Pareto's principle, you accomplish 80% of your output with 20% of your effort or, in other words, 80% of your production during 20% of your time.

However, the important message of the Pareto principle is that for familiar types of work, that is, in which one has a particular experience, a substantial amount of sound output can be generated in a relatively small amount of time. Giving priority to required, regular and maybe dull elements of the job ensures guaranteed output. In management, this is called "shortest processing time first." It frees your mind and generates quality time for the more challenging parts of your work. Examples of familiar work in realizing deliverables requested by your superiors are having your teaching courses for the students prepared in time, or performing a routine project for an external customer to generate extra income for your group.

### 4.3.4 Avoid wasting time

Of course, many external factors can disrupt your ideal work patterns entirely or use up disproportionate amounts of time. Examples are:
- Email and messages from social networks, especially if your mobile phone produces a sound each time a new message comes in. Our radical solution is to ignore nonprofessional social networks completely and pay minimal attention to research networks. Principally, the latter offer valuable services and help spread your work among the community; unfortunately, they regularly call for attention, sometimes several times per day. We try to check emails only a few times per day, we switch off notification services that beep every time a message comes in, and we often stow our phone away when we go to meetings or seminars.
- WhatsApp, WeChat, LinkedIn, and text messaging – all are great communication tools, but keep them under control; i.e., do not allow them to disturb. For example, inform colleagues and students that they cannot rely on these message serv-

ices to inform you or reach you. If the matter is urgent, use a phone call, or if it can wait, use email.

–   A continuous "open-door policy" can ruin your productivity. You do not want to be interrupted during your periods of quality time by colleagues, students, and salespeople who come by your office and expect your instantaneous availability. Realize that it takes 20 min to concentrate again on your work fully. However, you also want to be regularly available for your people. Practicing an open-door policy during certain times of the day only is a very effective solution for all. Quite often, essentials are exchanged and discussed in a few minutes, and everybody can go on with their work. It also prevents many meetings that regularly take an hour.

–   Meetings, especially periodical ones: The typical once-per-week management team meeting can be helpful to ensure that all members are on the same page concerning the organization. At the same time, it can also become a time waster when insufficiently prepared or when a standard agenda invites too much pondering on issues that do not need to be dealt with every time. As a rule, routine meetings where participants come unprepared are inefficient. Consider what needs to be discussed in the team and what is better handled face-to-face with individuals.

These remarks about wasting time at meetings are not meant to say that all meetings are always useless. On the contrary, they can be excellent sources of inspiration, chances to listen to others, exchange ideas, get people united behind common goals, and take jointly supported decisions. Limiting them in duration, however, is usually a good idea, and therefore, having a clock visible to all in any meeting room, and in your office, is a must. Just a 15-min coffee meeting to start the week and to inform each other on the current issues and priorities can work very well; at the same time, it is good for the social atmosphere and cohesion in your group.

### 4.3.5 Time management of coworkers: delegate small projects instead of tasks

An often hearsay frustration is that managers always seem to need more time, while their subordinates run out of things to do (Oncken & Wass 1974). For academics, this situation hardly occurs in the professor-student relation and, more generally, in the relationship between professionals with their own areas of responsibility and associated span of control. Nevertheless, it can play a role in the relationship with technical or administrative support staff. If you recognize this in your own environment, then consider whether your way of delegating tasks is optimal. It makes a difference whether you give assignments in concrete form of a to-do list (first do this, then that, report back when done, etc.) or that you delegate work as a project along with the purpose, the desired end result, and the appropriate responsibility to do what is needed. Completing such an assignment will generally give your coworker much more satisfaction than ticking off small tasks from a to-do list.

### 4.3.5.1 Briefing properly is key

Giving the appropriate information is essential when it comes to delegating work to your coworkers, support staff, and students. Much confusion and misunderstanding can result if this phase is underestimated. Briefing is key, and it takes time:

- If you delegate work to new or inexperienced people, always brief them in a personal meeting, monitor progress, and keep in touch at least daily. Remember that praise is always more effective than criticism, and starting off with praise gives better chances that the other will also be open to critical feedback.
- However, if work is delegated to an experienced coworker, briefing can be done in writing (email and memo), but personal contact is always preferable. Monitor progress as and when needed and agree beforehand on how to keep in touch on the project.

When delegating, it is essential to clearly explain the objectives of the work and why it is important. If persuasion is needed, facts and context work better than emotional arguments. Also consider whether the assigned work is a good match regarding the other's priorities and personal goals.

An important principle of delegating work is that one should never ask others to do something one would not be prepared to do; e.g., resolving a conflict at a higher level is something you should always do yourself. If it involves a coworker, one should solve it together, with the one who delegates – you – in the lead.

### 4.3.5.2 Upward delegation

Be careful to avoid that your people come back with partly finished assignments, hoping you will do the final part. If that is not what you agreed at the start, then do not accept incomplete work. It may be tempting to give in, as it may take you only little time to apply the finishing touch. The danger is that next time they try it again and return the job in an even earlier stage. Our recommendation is to be very strict: Once a task has been assigned along with proper briefing and transfer of responsibilities, it is your coworker's responsibility to complete it. In case of unforeseen difficulties or circumstances, you will be available for advice, but the completion of the task should not be delegated back to you.

### 4.3.6 Cultural aspects in the perception of time and the way it is spent

In the international setting of scientific research, one must be aware of the cultural background of team members. Significant differences regarding the concept of time and the way it is spent can exist. Some cultures place much value on spending long hours on the job, while in other countries, the emphasis is more on balancing time spent on work and on private life and doing work as effectively as possible. Working long hours is the norm in the USA and Asia, while in Europe, people focus more on a healthy work-life balance. Which approach leads to more success in the short and long term? We find it hard to say, although we know what we prefer.

We strongly favor allowing our coworkers flexible work hours, following the principle of working when it is most effective. For example, when a particular research facility in high demand is available, or a specific instrument is in top form, a motivated scientist will want to use it, no matter when it is after regular work hours (provided safety is guaranteed, of course). Enforcing strict work hours for motivated professionals is unnecessary. However, not all work cultures allow such flexibility.

Another cultural difference in this respect is the importance of being on time for appointments or obeying the time schedules of events. For example, for well-organized Western people, habits in other parts of the world that allow flexibility and improvisation, while sometimes disregarding appointments, can at first sight be disturbing and unsettling. However, our experience is that by being flexible where possible but using a bit of determination and persuasion to stick to essential items in a program, one almost always leaves with having the planned visits and meetings accomplished in the end, be it perhaps in a different order or timing than initially foreseen. And, always keep smiling!

### 4.3.6.1 Messy desks: effective or not?

Finally, time management literature generally emphasizes the importance of clean desks, simple file systems on your computer and in your file cabinet, and avoiding piles of documents and articles on the work table and random files on your computer desktop. We are not so convinced that this is always relevant. Obviously, one takes no risk with measurement outcomes and important data, even so with any document regarding financial administration, such as evidence of purchases and trips required for reimbursement from your own institution, or for being available when requested by grant agencies. We use a simple, systematic file system for anything that may be requested in the future, to avoid financial penalties or missed chances. However, we tend to be less systematic in filing articles, literature, and notes. Sometimes, these do pile up on our desks, amid notebooks and other stuff. "A messy desk is a sign of genius," is the saying, or according to Albert Einstein, "If a cluttered desk is a sign of a cluttered mind, of what, then, is an empty desk a sign?" A study by Vohs et al. (2013) titled "Physical order produces healthy choices, generosity and conventionality, whereas disorder produces creativity" gives confidence for not being overly concerned with the state of our desks.

## 4.4  A little project management

Scientists are generally not keen on project management. They often associate it with activities that can be planned, for which the outcome is known from the start, and with intermediate results that must be delivered at specific times ("milestones"). Indeed, in this sense, research as a venture into the unknown, with uncertain outcomes that can only be hoped for and are often entirely different than expected, is hard to approach as a traditional linear project. Nevertheless, many funding agencies require that proposals be written as a linear project, along with some form of planning. Ignoring this would mean we would lose a chance of funding.

For sure, many activities in academic or industrial research environments are suited for this type of planning and can very well be treated as linear projects. Just think about building a new laboratory or a new experimental setup, writing a publication, preparing a presentation for a conference, developing a new course or an entirely new curriculum, organizing a student excursion, a workshop, or a meeting, completing the final phase of a PhD thesis, etc. In such cases, linear project management can be of great value in accomplishing tasks in time.

Project management is highly developed and powerful (Nieto-Rodriguez 2021). However, a formal course can quickly go too deep for your needs, as it usually caters to more complex situations and is thus too versatile for straightforward projects. Hence, we believe it is helpful to present a short summary of linear project management here, as the "minimum you want to know." We assume that you will be the principal as well as the client who will use the project's outcome.

### 4.4.1  Definitions: projects differ from routine work

Let us start with proper definitions of the terms "project" and "project management." A project is a time-limited activity for achieving a unique and tangible end result (a product and a service), often defined in the form of deliverables along with specifications in terms of quality. The project has a defined beginning and end and usually a fixed budget. Note that the words "unique" and "time-limited" imply that a project is a special, one-off activity, has new content, and can never be a routine job. The management of such a project is therefore different from that of normal day-to-day activities. Project management encompasses the initiation, planning, organization, control, and monitoring of all the tasks and resources in the execution, required to achieve the intended result of the project, including the coordination of a project team, in case others are helping you.[1]

---

1  https://wiki.en.it-processmaps.com/

## 4.4.2 Scope, time, and resources

Three types of resources constrain your project: the **scope** (how broad you will allow the project to be, which aspects are included and which are not), the **time** in which it must be completed, and the **resources** (or budget) that are available, in terms of money and work hours of people in the project team or of external service providers. Cost is not always fully visible in academic environments, as time spent by academic personnel and support staff is often not accounted for (and we do not advocate that it should).

Defining the scope of a project for research, characterized by an unknown outcome, is not straightforward at all. Between the time while formulating a research proposal and the moment the research starts, the scope often must be adjusted, for example, because new knowledge has become available, or a new instrument has just been introduced on the market, or a new colleague contributes welcome expertise. Adjustment of scope will probably be needed several times during the project, with implications for timing and budget. Sometimes, this phenomenon is referred to as "scope creep," giving it a negative connotation, but in research, this is just a fact of life.

Example: The dean has instructed you to develop a short hands-on course on computer skills for first-year students (typical cohorts consist of 150 students) to consult the library, learn to use referencing software, and use graphical software packages for statistical data analysis. He wants it to be implemented in the first semester of the next academic year. He makes 160 h of IT support available and provides a budget of €35,000 for software and hardware.

**Project scope**: Introductory course in computer skills for 150 students, which they complete in 12 study hours
**Time**: Deadline for implementation is September 1 of next year.
**Money**: Budget of €35,000, 160 man-hours of IT support, and your own time.
**Quality**: Level to be determined with stakeholders

## 4.4.3 Stages in a project

The discipline of project management disposes over several different approaches and methodologies for various situations. We limit ourselves here to relatively simple straightforward projects, with stages as shown in Fig. 4.4: initiation, planning, execution, continuous monitoring and control, and closing.

Initiation: The project and its intended outcome are defined. A detailed analysis of the existing and new situations should be available, and the reasons for the desired change or for reaching the intended result should be crystal clear. Define the objective and scope of the project, the deliverables, the specifications, the costs, and the

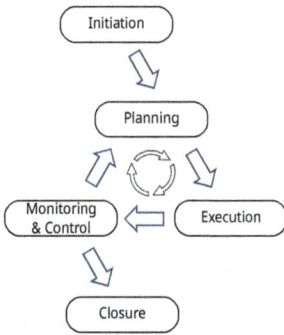

Fig. 4.4: The stages of a project. Note the planning-execution-control cycle, which can be traveled through for several rounds and gives the project an iterative nature.

moment the project should be ready. Check if the available budget is sufficient. A SWOT analysis as discussed in Chapter 5 may be appropriate to assess if the goals are realistic and the project is viable. The selection of a project team is also part of this stage, and for a very large project, possibly a steering committee is required. The initiation or definition phase of the project is very important and should lead to a clear and detailed project plan. Underestimating this stage may later on lead to serious problems and misunderstandings.

Planning: A detailed scheme is made of what needs to be done at what time; who has which responsibility; interference with the surroundings; people to be consulted or informed; what are the costs in terms of budget, time, manpower, and other resources; and an organization and governance structure including a communication plan (project meetings, how to report progress, and how to share information). A kick-off meeting is scheduled to make sure that all team members and other stakeholders understand the project, its reason, and the way it will be executed.

The obvious tool for planning the typical small-to-medium-sized projects in the academic world is the Gantt chart, a tool developed already in the early twentieth century by Henry Gantt. In its simplest form, it is a bar chart with time on the horizontal axis, indicating which phase of a project is executed at what time (see Fig. 4.5). More elaborate versions indicate the milestones and the interdependencies of subtasks, e.g., by colors or arrows. It is easily created with the help of a spreadsheet program, and many templates are available on the internet.

Execution: The project team carries out the various activities in accordance with the plan.

Monitoring and controlling: The almost continuous cycle of measuring the progress of the project, evaluating where it is supposed to be, deciding on how the activities and the plan should be corrected, along with keeping track of costs, hours spent, and adjustment of the plans. It is about keeping the project on track, on time, and within budget. It is also safeguarding the culture of the project: do the right people have the right roles and does the cooperation stay effective and efficient.

PROJECT X

| task 1 | | | | | milestone A | | | | | |
| task 2 | | | | | | | | | | |
| task 3 | | | | | | milestone B | | | | |
| task 4 | | | | | | | | milestone C | | |
| task 5 | | | | | | | | | | |
| task 6 | | | | | | | | | | |
| months | 1 | 2 | 3 | 4 | 5 | 6 | 7 | 8 | 9 | 10 |

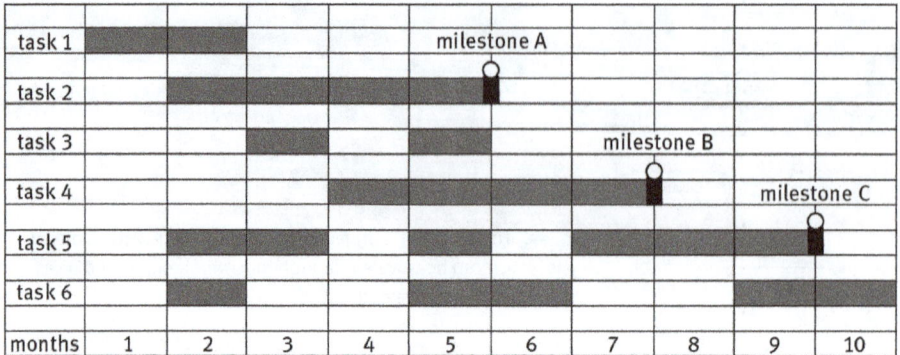

Fig. 4.5: The Gantt chart: a simple but useful tool to visualize the planning of a project.

Some grant agents request an audit upon completion of a project. Not having proper documentation on all financial matters or hours spent can involve penalties, implying that one does not obtain the entire budget. If you can afford a – perhaps part-time – trustworthy administrative assistant, you may be able to save a lot of time and frustration.

The three phases of execution, monitoring/control, and planning form a cycle through which the project iterates several times.

Closure: The final phase of a project implies ensuring that all project goals have been realized and the formal acceptance of the outcome by the principal. This phase also includes a report (administration, costs, bookkeeping, timesheets, logs, etc.) and an evaluation with the team to learn and celebrate. Maybe it is a good idea to convert the Gantt chart into a historical timeline, for example, by using the most important phases and achievements, as a graphical account of how the project was carried out. It may help you and others in future projects, and it can be a nice wall chart in your laboratory for showing visitors.

## 4.4.4 Scientists prefer agile management styles

As mentioned already a few times, researchers, especially those in academic environments, typically dislike the formal organization of project management. They are not the only branch. Since the turn of the century, agile management has come into fashion. This project management style is based on the Agile Manifesto (Beck et al. 2001), which highlights four values:
– Communication with stakeholders is more important than obeying standard procedures.

- Focus on delivering products and services that work and satisfy the customer's needs, and less on providing legally sound, and often voluminous manuals to document the outcome.
- Close collaboration with the client in all project phases is much more valuable than strictly obeying a contract.
- Openness to changes in scope during the project rather than rigorously sticking to the original plans.

Such ideas are generally close to the heart of a scientist, who is used to progressing insights and changing scopes versus budgets and resources that are typically constrained (Louttit 2024). Nevertheless, we do not propose to abandon all principles of project management, but we do like the flexibility and iterative possibilities of the agile approach.

## 4.5 Knowledge management and capturing implicit know-how

Researchers, educators, scientists, and developers, all of them deal with information, data, and insights – for simplicity, and probably not entirely correct, we catch it all under the term "knowledge." Organizing literature is conveniently done these days with so many sources on line and tools to build your personal databases and reference managers. Cloud services also greatly facilitate the storage of data and documents so that your entire team can access them and contribute. Not all information can be treated in this way, however. The management of knowledge in organizations has become an essential discipline since the 1990s, and multinational companies often have a chief knowledge officer in addition to the chief technology officer for technology and R&D (Koenig 2012).

Davenport's (1994) initial definition succinctly states: "Knowledge management is the process of capturing, distributing, and effectively using knowledge." However, there is more to it.

Knowledge comes essentially in three forms: *explicit* (knowledge set out in tangible form, i.e., which has been documented), *implicit* (knowledge in people's minds, which, however, can be made explicit), and *tacit* (knowledge in the form of experience or intuition that is almost impossible to document). An example of tacit knowledge is the "Fingerspitzengefühl" of a restaurant chef who knows exactly how to prepare the signature dish but would not be able to write it down precisely in a way that somebody else could reproduce it (and, probably, would not be keen to share it).

Recognition that knowledge comes in different forms is a fact in the nowadays generally accepted definition (Duhon 1998): "Knowledge management is a discipline that promotes an integrated approach to identifying, capturing, evaluating, retrieving, and sharing all of an enterprise's information assets. These assets may include databases, documents, policies, procedures, and previously uncaptured expertise and experience in individual workers."

We owe this comprehensive definition of knowledge management to Duhon, who, in 1998, published an article under the all-revealing title "It's all in our heads." That is exactly where a big concern of research leaders lies: How do we capture the knowledge of students and postdocs who worked with us for a few years when they leave? Publications and theses usually emphasize successful endeavors. Laboratory notebooks – provided your people fill these with the discipline that may be expected from professionals! – can give valuable extra information, and all data collected have probably been stored comprehensibly and are retrievable, maybe even the PDFs of articles referenced in theses and publications. However, much of the know-how leaves the laboratory along with people in the form of tacit knowledge. It is essential to realize this at an early stage. Succession planning can prevent the loss of implicit knowledge to some extent, when successors overlap in time with the people they succeed. Permanent technicians in the laboratory are also a great asset in this respect. Large research groups have better opportunities for such policies than smaller ones do, and the funding structure of universities or funding agencies does not generally allow effective succession planning.

### 4.5.1 Document essential procedures and "lessons learned"

To prevent essential knowledge and know-how from leaving your laboratory, consider implementing a strict policy for your students and coworkers to carefully document essential procedures, by writing a step-by-step manual on how to use an instrument, carry out a specific synthesis, or collect and analyze data. Such homemade manuals should be regularly updated, for example, with information on how to replace parts in a machine, how to solve specific problems, tricks in using the software, etc. Our strong advice is to do this from the installation and first use onward, even if the vendor provided a good manual. Such documentation is of tremendous value for new users, and it is an effective way to safeguard essential know-how. It is also good practice to capture essential know-how and knowledge at the end of each project, by organizing a "lessons-learned" meeting with all people involved.

## 4.6 Contact and relation management

Keeping track of important contacts is essential for researchers and academics. We want to remember not only our current students and alumni, past coworkers, and the peers with whom we interacted or may do so in the future. We also want to remember vendors, sales representatives, service engineers for our infrastructure, publishers, editors, funding agents, administrators, and science journalists, whom we may need to contact again. Many people will have moved on to new positions after a few years.

Hence, it is good to have a system that captures names and contact information, along with notes on why the contact was made.

The old way was to collect business cards, make notes on them, and keep them in a card index box or Rolodex system. However, nowadays, there are many software tools to help you. These range from the standard contacts function of your email system to programs such as Evernote, which let you digitize business cards and automatically complete the information by searching in online social platforms such as LinkedIn.

Maybe you do not need this. Perhaps you are the person who remembers every important meeting and conversation you had at a meeting. We, however, do not (maybe it is our age ☺), and therefore, we rely on simple forms of contact management. We strongly recommend that you adopt a contact management system at an early stage.

### 4.6.1  Relation management

One step higher than keeping track of contacts is what you do with them: relation management. Are you making use of your contacts and are you good at building mutually beneficial relationships in a network of peers, professionals, former students, friends, etc.? Do you keep the most important among them updated about your projects and progress? When you publish an important article, do you send PDFs to selected contacts with a short but personalized message? Again, having a consistent and adequate presence on social platforms can effectuate this to some extent. Some of our colleagues regularly publish a brief newsletter, for example, at the end of each year, and use it as Christmas greetings, via email or regular mail. It takes work, but it is an excellent way to keep your most essential relations informed.

Meeting face-to-face is nevertheless the most effective way to communicate. It is much easier to gauge what the other thinks in a personal conversation than by exchanging emails, and meeting in person, even if only short, is always the preferred way of establishing and maintaining good relations. However, in professional contacts, you must always be prepared to say the right things and to convey the right message. Sometimes, there is limited time to do so, so it is good to be prepared.

### 4.6.2  The elevator pitch: half a minute to get your message across

Imagine the following situations: By chance, you run into that influential scientist whom you always would have liked to meet, and you have no more than a minute before he must leave to get his taxi, what do you say? Or a radio reporter steps up to you at a conference, sticks a microphone under your nose, and asks your opinion on something. In fact, we discussed a similar situation already in Chapter 3, where we urged poster presenters to always have a 30-s summary prepared for visitors. For this type of occasion, you should always have a clear, succinct statement ready about what you want to communicate. It is often called the "elevator pitch," short enough to give a clear message to someone you meet by chance in the elevator, during a ride of less than a minute (although people usually don't talk in elevators).

You can think of an elevator pitch as a mini-speech to raise interest in what you and your research group do and to probe interest in a new idea, project, service, or strength you can offer. Doing it effectively takes preparation and experience; relying solely on your talent for improvisation is risky. Here are some recommendations for putting together a compelling pitch:

a) Define a clear objective – what do you want to achieve? A new collaboration? Finding a talented candidate for a vacant position? A joint application for funding? An invitation to visit and give a seminar at your colleague's institution? Access to a laboratory facility?

b) Include a brief explanation of what you and your group do, the subject you work on, the question you try to answer, or the problem you try to solve. Make sure it reflects your enthusiasm for the subject well. Stress your strong points (your "unique selling point") and why you and your group are an attractive party to team up with.

c) State your goal briefly and clearly, leaving no doubt that you have a clear plan of action. Remember, you are not merely asking for help; you have a proposition that can be of mutual interest (see Chapter 2: self-leaders have a plan).

d) If the pitch is used in a one-to-one contact, ensure you engage the other in some way, for example, by asking his/her opinion. If the pitch is for an audience, engage them with an open question or an invitation, and make clear how they can reach you if they want to follow up. In any case, try to avoid that the interaction is only a one-way transmission. Remember to present a business card or send an email to reconfirm your interest.

e) Exercise your pitch a few times, record it on video, and time it. The great advantage of going through this pain is that you will be much better prepared to give a concise message when the opportunity arises.

While it is good practice to always have a "message in a nutshell" ready in your mind for establishing a contact, it takes a step-by-step process to turn it into a sustainable, mutually beneficial relationship. We sometimes compare the process to a tugboat that will tow a supertanker. It takes a strong and heavy cable to do this, and getting it over in one go is impossible. Instead, contact is made with a much thinner rope, which is then used to pull a stronger one in, which, in turn, is used to get the ultimate cable used for towing. Building a lasting relationship works similarly; mutual trust and understanding develop gradually in steps. It breaks if the tugboat starts pulling full force

on the thin rope used to connect to the super tanker. Hence, developing mutually beneficial relationships takes time.

Chapter 6 discusses interactions and relationships between people in much more
detail and in a broader perspective.

## 4.7 Conflict management

Conflicts may arise in every situation where more than one person is involved.
Whether caused by a misunderstanding, a fundamental difference of views, or by
incompatibility of personalities, conflicts are simply a fact of life. They should be managed judiciously to avoid escalation and lasting damage to the functioning of your
team, organization, or even entire department or institute, or career.

Figure 4.6 summarizes the primary strategies for handling a conflict, whether between you and someone else or in your role as judge or moderator between members
of your team. It is wise to carefully adapt your strategy for resolving the conflict to
the situation. For example, in a threatening emergency, one should be forceful, take a
firm decision quickly and execute it immediately. You don't want to lose time in discussions, as the situation requires a clear and immediate resolution. However, as this
approach yields winners and losers, resolving every conflict this way is not wise.
Hence, it is only recommended for situations requiring immediate action or where
giving in would be disastrous for your interests or those of your organization.

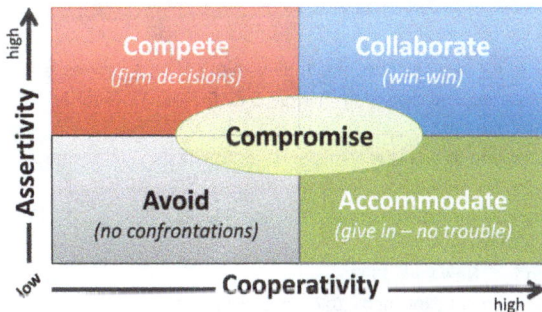

**Fig. 4.6:** Different ways of dealing with conflicts. Adapting the approach to the specific situation is
recommended.

Collaborating toward an outcome of mutual benefit – a win-win situation – is harder
to realize, but in the end much more rewarding. It implies that one should listen to
either side; explore opinions, feelings, emotions, and ideas of both sides; and mold
these into a resolution that is a win-win situation for all. One should probably reserve

this approach of handling conflicts for truly relevant issues of strategic importance, but not for minor things.

To accommodate as a conflict strategy means you give in to avoid trouble or persuade one of the parties to do so. This solution is applicable in cases of minor issues or if the losing party can be compensated by doing them a favor on an unrelated issue. The avoidance strategy tries to ignore conflicts or gain time in the hope that the problem will solve itself. The purpose is to avoid confrontation at all costs. This is sometimes possible when the source of the conflict can be removed, e.g., by appointing a new employee or changing directions in an ongoing process.

Finally, one can negotiate a deal between the parties in the conflict. If both sides are willing to move a bit, reaching a compromise that both sides can live with may be possible.

We recommend that you reflect a little on the strategies for dealing with conflicts and try to avoid applying one approach in all cases. It is interesting to make the connection between conflict handling strategies in Fig. 4.6 and the personal operational dimensions in Fig. 2.1. The drive dimension obviously relates to competing and using personal power, while the feel and do dimensions are probably more connected to less assertive strategies of avoidance and accommodation. The doer might want to avoid and continue. The feeler might prefer to accommodate. The thinker might prefer the compromise approach. A combination of the driver and the feeler might prefer the win-win strategy.

## References

Beck, K, Beedle M, van Bennekum A, Cockburn A, Fowler M, Grenning J, Highsmith J, Hunt A, Jeffries R, Kern J, Marick B, Martin R, Mellor S, Schwaber K, Sutherland J, Thomas D (2001) Manifesto for Agile Software Development. http://agilemanifesto.org/.

Cohen, C.M., Cohen, S.L.: Lab Dynamics, Management and Leadership Skills for Scientists; Cold Spring Harbor Laboratory Press, 3rd Edition; Cold Spring Harbor, New York, 2017.

Davenport, T.H.: Saving IT's soul: Human centered information management. Harv Bus Rev; 1994; 72; 119–131.

Duhon, B.: It's all-in our heads. Inform; 1998; 12; 8–13.

Hindle, T.: Manage Your Time; DK Publishing Inc., New York, 1998.

Koenig, M.E.D.: What is KM? Knowledge Management Explained, 2012. http://www.kmworld.com.

Lebrun, J.-L., Lebrun, J.: The Grant Writing and Crowdfunding Guide; World Scientific, Singapore, 2017.

Louttit, J.: Leading Impactful Teams, Achieving Low-Stress Success in Project Management; De Gruyter, Berlin, 2024.

Manifesto for Agile Software Development: 2001, www.agilealliance.org.

Nieto-Rodriguez, A.: How to Launch, Lead, and Sponsor Successful Projects; Harvard Business Review Handbooks, Boston, 2021.

Oncken Jr., W., Wass, D.L.: Management time: Who's got the monkey. In: On managing yourself. Harv Bus Press; 1974; 52(6); 75–80.

Vohs, K.D., Redden, J.P., Rahinel, R.: Physical order produces healthy choices, generosity and conventionality, whereas disorder produces creativity. Psychol Sci; 2013; 24; 1860–1867.

# Guest column: Time, money, and great ideas

Graham Hutchings

*Graham Hutchings is Regius professor of chemistry at Cardiff University.*

My best advice to young academics setting up their own research program is to focus on doing high-quality research. Choosing the right field and problems to work on is essential. You want to ensure that you are not doing the same as everyone else. Recognize that great research takes time, success is not assured, and problem selection is key.

For me, scientific leadership boils down to three key points: great ideas, money, and time.

- Successful research on unique projects has very low odds, you will need lots of ideas to succeed, and hence, you need to constantly generate new ideas. Be ruthless with ideas and select the ones you think are best and will create impact.
- Money: Chase it! To me, there is always the thrill of the hunt. Make sure you are aware of the opportunities and learn how to write convincing proposals; this is an essential skill for a successful researcher. Networking and engaging in collaborations are very good and recommended, but collaborate with your peers, and not with celebrities, as these are likely to get most of the credit. Always make sure that your unique contribution will be visible. In an application, use a track record section to sell yourself, and put in key results that underpin the application.
- Time is the most important commodity that you have; use it wisely, and ensure that you have quality time for new ideas and for thinking deeply about your ongoing projects. Time management is another essential skill; learn it early and apply it. It is essential for being focused, and to avoid that you spread your precious time over lots of outcomes. Involve your team as actively as you can, in writing papers and proposals, submitting abstracts for conferences, and in teaching. They help you, and they learn a lot themselves too.

## Publication strategy: Think about a research portfolio

You should also think in an early stage about your publication strategy. If you want to aim for the highest-impact publications, do *not* pre-publish in any way; once you have shared your ideas at a conference, your peers or referees will have a head start. Nevertheless, in the present academic climate, you will have to publish to build and maintain a reputation. The consequence is that you will need a *portfolio* of research, with projects that generate output that can be published rapidly and a smaller fraction of well-chosen high-risk-high-impact research.

## Make sure you and your work are noticed

Try to position yourself and your students for selection as a prize winner. This can be at the beginner's level with the best presentation or poster award or winning a travel grant, or once you are more experienced, you may try to go for prizes for your research or publications. In other words, get yourself and your work noticed.

## Personal lessons

I end with a few lessons learnt in industry and academia.
- Recognize when you are lucky – some of the most important discoveries are made when an unexpected result is obtained that challenges rather than confirms the hypothesis being tested. Remember "chance favours only the prepared mind" (Pasteur).
- Learn to manage your manager, and if you don't like the system, change it! This was advice I was given by my first boss when I joined ICI (Imperial Chemical Industries) – mind you, he was both very successful and unmanageable.
- How do you make sure you get the time to make that great breakthrough? Once again, time management is key.
- In your team, ensure that it is clear "Who has the monkey" – I learnt this in a management course while in AECI (African Explosives and Chemical Industries). Think of a task as a monkey that sits on the back of the person who has to get the job done. Make sure that you don't allow the upward delegation of problems and partly completed assignments, because then, you will end up with all the monkeys' back with you on a Friday afternoon. If this happens, it can cause you a lot of your valuable time to be wasted – don't allow it. In your team, educate your people to be problem solvers, and not problem creators. So if you are managing a team, make sure you control the workflow to and from you!
- Leadership is about seeing all sides and not asking your team to do something that you yourself are not prepared to do.

Finally, take a step back now and then to reflect a little on what your aims are. Are you heading for a future in academia or in industry? Are your work and personal life in balance? Considering these things is important early on. Overall, remember that research is fun and being the leader of a research team is a great privilege. Finding out something new is exciting and can lead to great rewards for both you and society. But if it isn't fun, it's not worth the effort in my view.

# 5 Planning the road to the future: strategy and comprehensive plans

## 5.1 Introduction

To realize your ambitions, you wish and hope that your research group, laboratory, and institute operate like a team that wholeheartedly supports your vision and that things automatically go in the right direction. This takes optimal decisions about everything they have to do and operates in a way that enhances each other's performance. This would be the ideal situation, and if you succeed in getting anywhere near, you can consider yourself extremely lucky. What kind of leadership would be needed to form a self-organizing, synergetic team?

Effective leadership involves your coworkers in a way that they work for their own and your goals. To let this happen, these common goals must be clear to all team members, including why these are important. These goals derive from, or at least fit in, a shared vision, a mission, and a strategy for what the team wants to achieve as a whole and a vision, mission, and strategy for each team member separately. In general, your research group will have to fit into an organization, e.g., the university department you belong to. Hence, you must know what your superiors have in mind.

## 5.2 Vision

First of all, you need a vision that you and your most important coworkers passionately believe in, perhaps not immediately, but in any case after some time – a vision that is hopefully compelling to all people in your team and shared at all levels, a vision also that complies with the view of the larger organization you belong to (e.g., your university, or a multinational company, or one of its divisions). The vision describes the way you see your organization in a larger context, your research group, or yourself after you have reached your goals. It is the optimal desired state of the larger environment to which you belong (your department, institute, university, maybe your discipline, or even your country) in the future, say 10–25 years away from now. You could also call it a dream, but it should be realistic. A vision inspires all activities and leaves room for creative interpretation and new ideas. This vision generally has a considerably larger span than what you want to achieve with your organization in the coming years.

**Example of a vision**
Suppose you are working in the sustainable energy field. You would then probably have a vision of what your country or part of the world would need to do to reach a sustainable energy system in the next few decades. Your vision might be that solar and wind energy parks replace all fossil-fuel power

https://doi.org/10.1515/9783111325644-005

plants in the next 10 years and that managing such a transition would need significant technological development in material improvements for cheap and reliable solar cells, mechanical improvement of windmills, development of batteries for electric vehicles of all sorts, electricity grid management systems for optimal distribution of electricity, and storage systems for excess electricity on a super sunny or windy day. In addition, the world would need a trained workforce at many levels and in many disciplines, and the educational implications will be substantial. It would take decades and many research and development groups, organizations, industries, and governments to get this all done. Such a dream and overarching plan could be called a "grand vision."

You can make only a small – though undoubtedly meaningful – contribution. Your vision should also describe this. Your research group would concentrate on one important aspect. Let us say the development of a new type of solar cell or anode of a battery, and you have clear ideas of what the technical bottlenecks are and how the associated problems could be solved, based on a novel combination of materials. This could be your vision.

It is essential to realize that your research group is part of a university department or an industrial R&D laboratory, perhaps of a multinational company. The larger organization to which you belong will also have a vision, and it is essential that you know it thoroughly and consider how your vision relates to it. On the other hand, your students, coworkers, and staff members may also have a view on how they see their roles and their future. Ideally, these visions, although probably quite different in scope, should match and not be contradictory.

| The vision | | |
| --- | --- | --- |
| **The organization** | **Your unit** | **Yourself** |
| How does the organization see its role in society, the world, and the sector it is active? | How does your unit fit into the organization and the sector in which the organization is active? | What is or should be your place in the organization/unit where you will be employed? |
| – What is the spirit that drives the organization? | – What is the spirit that drives your unit? | – What is the spirit that drives you? |
| – In which "market" or "sector" will the organization operate? | – In which field will the unit be active? | – What are your prime motives? |
| – Which role does the organization want to fulfill and how does it want to distinguish itself? | – Which role would the unit like to fulfill and how can it distinguish itself? | – How would you like to live and work? |
| | – What are the unit's prime motives? | – In which field would you like to work? |
| | | – Which role would you like to have? |
| – What are the organization's prime motives? | – What kind of working and living culture would the unit like to have? | – How can you make a difference in your expertise? |
| – What kind of working and living culture would the organization like to have? | – What energizes the unit? | – What energizes you? |
| – What energizes the organization? | | |

The scheme below lists several questions that will help you formulate a vision for your unit in relation to the organization it belongs to and for yourself personally. Once you present your views on the future, your team members may use the right part of the scheme to develop a personal vision for themselves.

## 5.3 Mission

Where your vision statement explains how you ideally see your organization's role in a greater context, your mission statement describes what you want to achieve in the coming years. In this sense, it is more down-to-earth and tangible than the more idealistic grand picture. The mission describes the current intentions of your organization. It specifies what it does, for whom and how, with a focus on a shorter term of 3–5 years from now. The following scheme may help you formulate an effective mission statement.

**Example of a mission**
Continuing with the vision for research on solar cells in the sustainable energy field as described above, the mission could, for example, be that you wish to explore the applicability of a new complex material that you anticipate to have excellent properties for converting solar light into electricity and to optimize its material properties for this purpose. After demonstrating the suitability of the material and further improving its properties, you may want to build a demonstration cell to prove that the principle works in practice and that application on a larger scale is viable. In parallel, you may want to work on a preparation route that enables scaling up and establish contacts with industry to exploit the ideas further and perhaps work toward applications. You estimate that the realization of this mission should be possible in 5–7 years. The mission statement could be: "Explore new materials for efficient conversion of sunlight into electricity and establish contacts with industry to explore their application."

Note the difference in nature between the idealistic vision and the more practical mission statement. Our own experience in academia is that many prospective students looking for a research group where they want to graduate are often more attracted by a clear picture of what the professor and his/her research group intend to do in the near term than by the bigger picture. This is logical because they generally aim to finish in time and with a successful thesis. Hence, having both vision and mission statements in order is essential.

Vision and mission together clearly describe what you want to achieve. It may be that your coworkers or students are more interested in your mission than your vision, or even in some concrete goals that derive from it, as it has immediate impact on their activities now. With a crisp and clear mission statement, you may attract new students. However, the organization you belong to, including your head of department or even the executive level of your university or research institute, will have a keen interest in both, especially where you intend to go in the long run.

| The Mission | | |
|---|---|---|
| **The organization** | **Your unit** | **Yourself** |
| How does the organization see its role in the society, the world, and the sector in which it is active? | How does your unit fit into the organization and the sector in which the organization is active? | What is or should be your place in the organization/unit where you will be employed? |
| – What does the organization want to contribute? | – What can your unit contribute? | – What can you contribute to the organization/unit where you will be employed |
| – What does the organization want to accomplish? | – What would the unit like to accomplish? | – What do you want to accomplish? |
| – Does the organization have a clear mission? | – Does the unit have a mission of its own? | – Do you have a mission for yourself? |

## 5.4 Goals for short, medium, and long term

The next step is to derive some concrete goals from your vision and mission, for the short term (say 1–2 years), the medium term of a few years (in research, typically the time required to finish a PhD project), and the long term, say 5–10 years from now (see Fig. 5.1). Don't be afraid to formulate these. They are not cast in stone, and you can adjust or even radically change some of the goals where necessary.

**Alternative views: what comes first – vision, mission, or goals?**
The way we have described vision, mission, and goals may suggest that formulating them is a linear process over time. This does not have to be the case. Many people may start from a mission of what they want to achieve, or even a rather detailed set of goals, while a vision comes later. Different views on strategy development exist. Mintzberg (1990), for example, distinguishes 10 different schools of thought on strategy development. In the Anglo-Saxon world, one often starts from a mission and a set of goals, while the vision is seen as a view on how to achieve these, and the term "strategy" refers to the plan for realizing the mission. Alternatively, strategy development in continental Europe – sometimes referred to as the Rhineland approach – regards the vision as an idealistic dream, from which mission and goals follow. Here the term "strategy" refers to the entire process from vision to realization. In comparison, the Anglo-Saxon approach is to the point and aims at tangible results, while the Rhineland approach strives for comprehensiveness.

## 5.5 Strategy

For implementing the vision, we need an integral plan of action that takes all important aspects into account – a strategy for achieving your goals.

Fig. 5.1: Vision, mission, and goals for the short, medium, and long term need continuous revision, and are dynamic.

The concept of "strategy" may have different meanings and can be defined in several ways. Mintzberg et al. (1998) distinguish between several types, of which we mention three here:

- Strategy as a plan – a directed course of action to achieve an intended set of goals.
- Strategy as a pattern – a consistent pattern of past behavior, with a strategy realized over time rather than planned or intended. Where the realized pattern was different from the intent, Mintzberg refers to the strategy as emergent.
- Strategy as a position – locating brands, products, or companies within the market, based on the conceptual framework of consumers or other stakeholders. This is a strategy determined primarily by factors outside the firm.

In scientific research and development, one can undoubtedly recognize these three variations. In particular, the concept of emergent strategy will look familiar to many scientists. For example, one can make a beautiful research plan to investigate a specific burning scientific question. However, suppose one cannot obtain the funding to carry it out in its original form. In that case, one has to redefine the plan in a somewhat different direction to benefit from another funding scheme. The increasing scarcity of funds and diminishing success rates of grant proposals have urged many researchers to adapt strategies in an agile manner – a typical case of working with dynamic and emergent strategies. *Intended* strategies adapted based on *emerging* strategy make strategies dynamic and not cast in stone, see Fig. 5.2. We will return to emergent and dynamic strategies in Chapter 7 (Fig. 7.5). For the moment, we limit ourselves to the concept of strategy as a plan, with

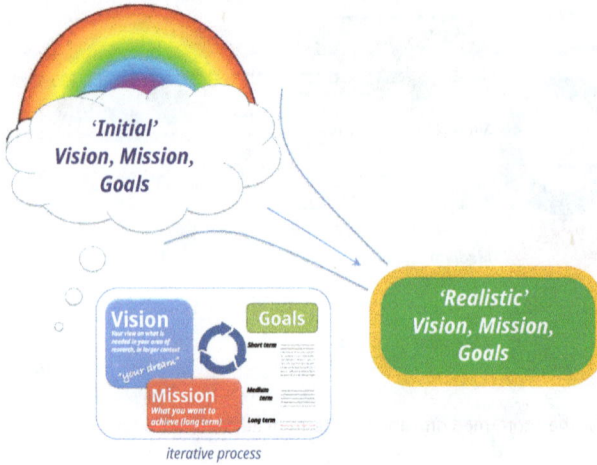

**Fig. 5.2:** Over time, too idealistic dreams reduce to a feasible vision, mission, and goals, in line with one's span of control.

building up a new research group or perhaps a new institute as the primary example in mind. So, for us, strategy is a plan of action, a translation of vision and mission in a plan to achieve your goals. Before we formulate it, we need to consider a few essential points.

### 5.5.1 The playing field: explicit and implicit rules of the organization

In any sport, one needs to know the rules of the game before one can win a match. The game will have explicit rules, written out in a book, as well as implicit or unwritten rules, relating to culture, habits, or spontaneously grown customs (e.g., in soccer, to kick the ball out of the field if someone from the opposite side gets hurt; it is not in the rules but every team does it). Also, in science, one should be well aware of the playing fields, which are the set of conditions, constraints, opportunities, and rules of the organizations in which we work or deal with, under which we will try to realize our mission. An essential part of this is that we understand very well how success is defined for yourself and for the organization (e.g., your department, university, and company) to which you belong (see Chapter 1). Every organization has its own collection of explicit and implicit rules. Some you may be able to challenge or even ignore, but most, you will have to obey and live with. Being aware of these can save you some frustration.

### 5.5.2 Strengths, weaknesses, opportunities, and threats: the SWOT analysis

It is essential to make a realistic assessment of your position, i.e., where your strengths and weaknesses lie, and how these combine with the opportunities or threats in your area or environment. This is the famous SWOT (strengths, weaknesses, opportunities, and threats) analysis, which is a generally accepted assessment tool in the world. The big advantage of the tool is that it does not rely on any theory and represents an objective, holistic method to evaluate one's position or chances. It applies to an individual, a group of people, an organization, or even an entire country. Figure 5.3 shows a SWOT diagram. The fields "strengths" and "weaknesses" are internal factors and can, for example, be affected by changing the composition of a team or finding strong collaborators. Opportunities and threats are always external factors, outside one's span of control, dictated by the outside world. However, they can also be influenced by one's own initiatives and activities. The appendix has a worksheet that can be used as a template for your own SWOT analysis.

---

**Failed research proposal due to violation of the SWOT analysis**

It sounds almost silly in Fig. 5.3 to assume that one would venture into an opportunity without having particular strengths, but it happens often in science. A certain professor tried several times in his career to obtain grants outside his specific field of expertise. In 2010, funding opportunities were dwindling in his country, until a fairly substantial funding scheme was announced for developing the chemistry of converting natural sources such as lignin and cellulose from plants into base chemicals and fuels. As a materials chemist, well-versed in solid-state aspects of matter but not in organic reactions, he submitted a proposal on developing new catalysts. It was rated "good and worthy of support when funds are sufficient" but not as "excellent" or even "very good," which was the threshold for granting due to oversubscription of the program. This is a typical example of going against the SWOT analysis principles. Had he teamed up with a specialist in organic reactions, he might have had a better chance.

---

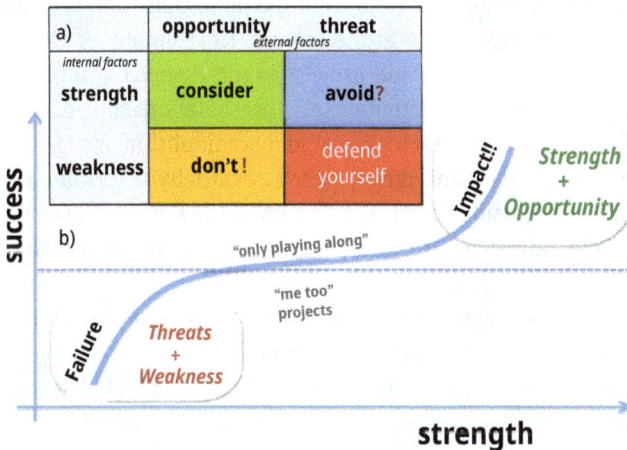

**Fig. 5.3:** How to base an action plan on the outcome of a SWOT analysis. Combining your strengths with favorable opportunities is your best bet while you wish to avoid areas where threats might harm you extra on your less strong points.

### 5.5.3 A comprehensive plan

After formulating a vision, a mission, goals, and a strategy, possibly tested with a SWOT analysis, it is time to formalize all this in a comprehensive plan. The scheme in Fig. 5.4 offers a tremendously valuable tool for planning, as well as for analyzing where you stand at any given moment. It is available in the appendix.

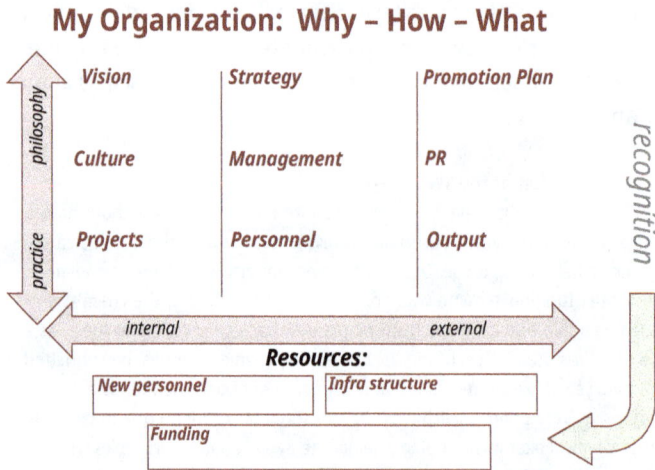

Fig. 5.4: A comprehensive plan addresses all essential aspects of your organization.

The scheme is built on two axes: vertically, it runs from the philosophy behind your strategy at the top to the practical realization at the bottom. Horizontally, it ranges from internal matters such as vision, culture, and projects to the external visibility of your organization through public relations activities (e.g., presentations at conferences, newspaper interviews, and public outreach events) and accomplishments (graduated students, articles, patents, posters, and talks), which eventually determine the recognition for your group by the outside world. The bottom of the scheme shows the enablers for your work: people, infrastructure, and funds. We will go through each of the headings separately:

- **Vision**: A short version of your vision, mission, or even goals (as you prefer). This part has its influence on all other elements of the scheme.
- **Culture**: The core values of your organization, as fundament for thinking, behavior, and cooperation in your research group, department, or organization. Examples of core values are safety, ethics and integrity, open exchange of ideas, and mutual respect.

- **Projects**: The way your vision on, e.g., research and education translates into practice. These entail specific research projects, the courses you teach, perhaps also initiatives to structure your discipline on a national or even larger scale.
- **Strategy**: A concise description of how you want to realize your vision/mission/goals.
- **Management**: The way you structure the necessary activities in your unit, such as teaching and supervising students, coordinating safety procedures, infrastructure, and supplies, funding, reporting, general management, and regular meetings.
- **Organization:** Structuring and organizing the necessary activities in your group, including accountability and responsibilities. This may include a short description of necessary procedures for your group (a mini *Soldier's Handbook*); for example, how to arrange project and progress meetings, periodic seminars, management team meetings, procedures for safety checks of your laboratories, budgeting, reporting, and monitoring literature. Please keep it simple and preferably as lean as possible. However, ensure that critical aspects are appropriately arranged.
- **Promotion plan:** Your strategy for letting the world know what you are doing, how successful you are, and to persuade your network to recognize, promote, support, fund, or buy from you or help you in any other way. Contact and relations management (Chapter 4) is critical here. You identify target audiences (stakeholders such as industries, companies, funding agencies, professional societies, your peers in your discipline, and the students you would like to reach and attract to your group) and why and how you want to inform them with what sort of frequency. Please realize that just publishing scientific papers or giving the occasional talk at a conference may not be enough to distinguish yourself (see also Chapter 3 on communication and presenting science).
- **Public relations**: A concrete scheme for informing different stakeholders of your intentions and successes, involving them and building an attractive image around the identity you and your group want to be. Examples are an attractive and informative website that is regularly updated, presence on selected social media, regular visits to companies, twice-per-year popular article for the larger public (to appear in a local or national newspaper or a magazine of the scientific community), and a press communiqué to publicize a genuinely remarkable finding or the fact that you managed to publish a paper accepted in a highly ranked journal. Also, think about a policy for keeping students informed about your activities to stimulate the best ones to choose your group for their final projects or your department for their graduate studies. Being present on social media or having short video highlights of your group's activities can also be very effective. Finally, having a plan for when and where you or your coworkers and students present at conferences is essential for promoting your work in your field's community, obtaining feedback and new ideas and making new contacts.
- **Output and results**: Here, you probably think first of the scientific papers and reports that you publish, the patent applications that you file, the designs you present at an exhibition, the posters and presentations by you or your students at

conferences, and the abstracts in the program book. However, also the students who graduate from your group or your department, or the postdocs who leave for academic or industrial positions count as most valuable accomplishments of your activities.

– **People, infrastructure, and funding**: At the bottom of the scheme are your intended resources, where and how you plan to recruit your staff and attract your students, how you want to build up your laboratory infrastructure, and finally, where the funding may (or will) come from.

The upper part of the comprehensive plan in Fig. 5.4 has three columns: the left one for the internal aspects (vision/mission, culture, and projects), the middle for operations (strategy, management structure, and daily organization), and the right column for your external visibility. It is the latter that will lead to recognition for your endeavors. Therefore, it is of great importance for your chances to obtain funding for excellent infrastructure and top-quality people who like to work with you and also for giving you a good position for valuable collaborations that will widen your opportunities to be successful in science.

### 5.5.4 A procedure for assessing your research unit, department, and organization together with your coworkers

The scheme in Fig. 5.4 can well be used to reflect on your present activities, either by yourself or, even better, with some of your coworkers. We describe an effective procedure that both authors have used many times in assessments of all kinds of organizations, ranging from small research groups to prominent governmental research institutes and a variety of large companies. The procedure works as follows:

1)  Invite some of your team members to a half-day meeting, preferably "off-site" with mobile phones switched off, to minimize distractions. It works best if you involve members from different levels, e.g., for a small research group, perhaps everybody, and for a typical European academic research group, the permanent scientific staff, a technician, the secretary of administrative assistant, and one or more students.

2)  You will need a pile of sticky notes in three colors – green, yellow, and red, for example – and enough felt pens and a poster-sized version of the "My Organization" scheme of Fig. 5.4. Ask the participants to take 15 min to think about the organization and its activities and write down what issues, problems, and successes that come to their minds: one subject per sticky note, green if it counts as successful, yellow if it needs attention, and red if it goes wrong. Anything that comes to mind is good, on every level, be it a daily problem in the laboratory, an administrative hick-up, a successful appearance in a conference, the failure of a grant proposal, student complaints about a challenging exam, a perceived short-

coming in the vision behind your organization, everything that comes up is essential, no matter at what scale or level. Figure 5.5 gives an idea.

3) When all have finished writing, the notes are collected, briefly discussed one by one, and each note is placed on a poster version of Fig. 5.4, in the category where it belongs. People cannot withdraw notes but may add new ones, stimulated by the discussion. It is important that you, as a leader of the discussion, do not respond too much or start to defend yourself. The aim is to get all issues your team members find important on the table. Whether negative or positive, fair or not, the point is deemed relevant enough to be brought up by your team members, so you must appreciate it as necessary. Also see it as a comment on the organization rather than on you.

4) At the end of the discussion, you will have the "My Organization Poster" clotted with colored notes, perhaps looking as the example in Fig. 5.6. What does it tell you? First, it shows where the team's focus is and whether you are paying sufficient attention to all the important aspects of running your organization. Figure 5.6, for example, appears to reflect a team of doers and drivers with a strong focus on projects and output. The green notes indicate that most of projects are going well, and the predominance of green on the external side is evidence that the group can boast some successes. The culture is evidently pleasant and inspiring as the positive feelings on the team's output indicate. However, the daily organization and management structures seem to lack structure. Perhaps, the group leader relies too much on improvisation or allows urgency as the main driver here. Vision and strategy clearly need more attention, as the team does not seem very aware. Of course, it is always possible that the leader has the vision, mission, and strategy clearly in mind, but sharing it more with the members may be good and a first step toward getting the daily organization in better shape. Also note that a strategy for public relations is mostly missing and that one or two persons raised concerns whether the team receives adequate recognition for its endeavors. The resources clearly need attention, with the acquisition of funding as the most critical concern.

## 5.6 Concluding remarks on vision, mission, and strategy

To summarize this chapter, as an accomplished self-leader and as the leader of an organization, you must have a compelling and well-articulated vision and mission.

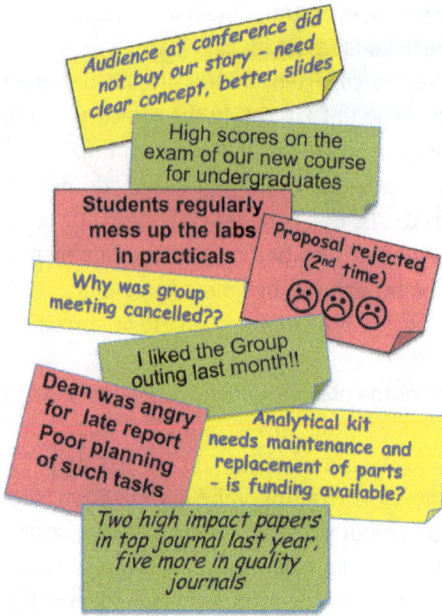

**Fig. 5.5:** Each note mentions one issue, with colors green, yellow, and red indicating success, needing attention, or failure, respectively. In a real situation, one typically has 50–100 of such notes.

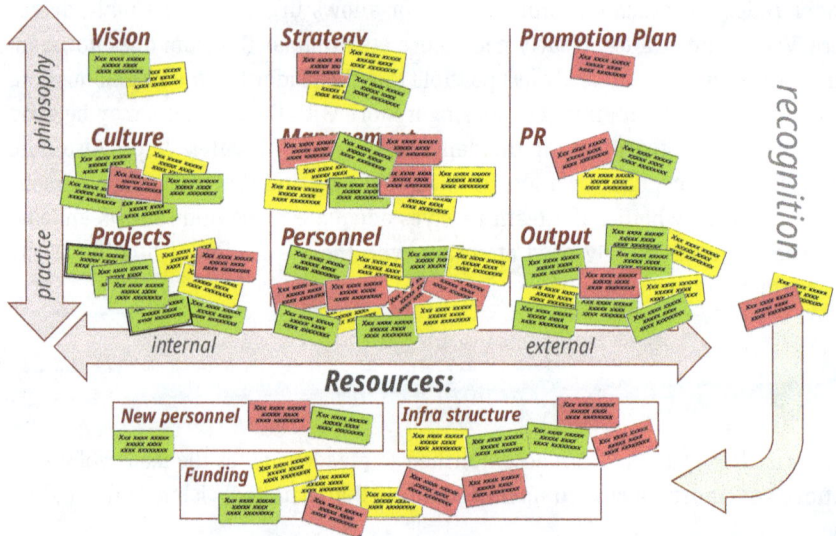

**Fig. 5.6:** The team meeting leads to an overview of all issues shared by the team. The distribution clearly indicates where the group's focus is and which aspects need more attention.

While the vision reflects your dreams and goals in the longer term, your coworkers (students) may be more interested in your mission statement and a concrete set of plans and objectives for the coming years. The strategy is a plan of action on accomplishing your goals, with a comprehensive view on internal aspects (vision, culture, and project teams), daily operations (tactics, management, organization, and ongoing projects), and how to present your achievements to the world. The latter is crucial for your chances to obtain the necessary resources (funds, infrastructure, and people) to realize your dreams.

## References

Mintzberg, H.: Strategy Formation: Schools of Thought. In: Perspectives on Strategic Management; Fredrickson, J. ed. Harper & Collins, 1990.

Mintzberg, H., Ahlstrand, B., Lampel, J.: Strategy Safari: A Guided Tour through the Wilds of Strategic Management; The Free Press, New York, 1998.

# Guest column: Science is the right guide for the future

Yong-Wang Li

*Prof. Li is Founding Manager of Synfuels China Technology Co., Ltd., in Beijing-Huairou.*

The climate for high-quality scientific research is not so good at the moment, I'm afraid. That is a pity, because we should all be working hard on creating solutions for a safe and sustainable future for our world. In the energy sector, for example, without fundamental scientific knowledge, we will not be able to find solutions that are truly durable. In the end, I think we will have to look at properties and behavior of materials at the yet hardly explored quantum level, to discover future energy resources that are related to the fine control of the fundamental structures and functionalities of matter due to weak interactions that we are now unfamiliar with. It easily takes 50–100 years to discover such knowledge and transfer it into technology.

For companies, such new fundamental knowledge will help us in safeguarding our competitive position, but we must be patient and we need to have courage to take a long-time perspective. And we – including our shareholders – must also be prepared to invest the money we make today in definitive solutions for the far future.

It may sound contradictory, but I strongly believe that meanwhile, we must continue to use energy from fossil sources. The clean and responsible application of such fossil fuels can give us time for working on real solutions. It also generates funding for new scientific know-how and technologies and for educating young scientists for these tough challenges. We should encourage our younger generation to work in these essential and new areas.

Unfortunately, other forces exist, some push forward, but some hold us back also; many artificial regulations, outdated policies, or stupid power games have a negative influence on progress. The funding climate for academic research makes it difficult to work on long-term goals, and the too competitive climate at universities stimulates fragmentation: small projects, fast publications, and short-term successes.

In my own trust, science always gives us the right guide to the future. It is so important that we scientists take our responsibility to explain what we do, why we do it, and why we want to devote our precious time to it. The true driving force should come purely from science. In the end, it makes no sense to distinguish between pure and applied science – science is there to solve problems.

# 6 Understanding how to build successful teams: creating synergy between people who trust each other

## 6.1 Modern science is teamwork

Modern research and development is teamwork. Unlike a century ago, one can hardly do anything alone. It is not at all uncommon to see scientific articles published by 10–20 authors, belonging to more than one academic laboratory, and often at different universities as well. In our own research, although positioned at the fundamental end of the spectrum, we have often fruitfully collaborated with research departments of large chemical companies, such as Shell in Amsterdam, SASOL in South Africa, and Synfuels China Technology in Beijing. Industrial researchers usually know the real problems and questions from experience, but often they cannot afford to delve into the fundamental aspects underneath. This is where academic research groups have their strengths. On the other hand, our industrial colleagues can tell us whether our efforts in research make sense or whether we are trying to find solutions for irrelevant problems. As our distinguished mentor Dr. Jens Rostrup-Nielsen (2016) often warned us: "University professors should beware of non-applicable applied research."

What makes a team successful? In any case, the members should share an interest in the same goals and believe in similar values, among which trust, honesty, modesty, and discipline are the most important. Moreover, the members bring talents, expertise, and know-how that are mostly complementary, with perhaps some overlap to enable mutual understanding. Finally, effective leadership plays an important role, although in self-propelling teams, the members lead intermittently, as the situation demands. We then talk about shared leadership.

According to Lencioni (2002), members of a cohesive team exhibiting positive synergy behave as follows:

1. They trust each other.
2. They openly discuss ideas critically in a constructive way.
3. They commit to decisions and plans of action.
4. They hold one another accountable for not delivering according to those plans.
5. They focus on achieving collective results.

Points 2 to 5 of Lencioni regard tangible aspects that are very important for producing concrete results. Point 1 refers to the intangible element of trust. People have within them a lot of hidden potential in the form of talents and thrive, as well as willingness to connect and share. Being aware of this, they can lead themselves from within with dedication and inspiration to make their specific contributions. Also a productive process of cocreation and intercreation may start. Suppose a leader succeeds in combin-

https://doi.org/10.1515/9783111325644-006

ing the intangible value of human sources with the tangible process of creating results in a team. In that case, it can reach high levels of performance and impact.

Therefore, a team, whether it consists of your students and postdocs or your experienced colleagues, will have better chances of success when the members feel that they are entirely accepted as they are. Thus, they think that they belong to the group and are valued for their contributions. This is the so-called 3B principle: **be** oneself, **be**long to the group, and **be** valued, as we discuss later in this book.

## 6.1.1 Hierarchy in teams

Self-propelling teams change leaders all the time, sometimes many times per day, and perhaps even without noticing it. This principle is well illustrated by Oshry's (2007) attractive team model entitled "total system power" shown in Fig. 6.1. The team members, irrespective of their formal position or function in the organization, can have varying roles, depending on the situation. These roles can be classified as top, middle, bottom, or customer, which can be seen as conditions that all members face, irrespective of their formal position or function in the organization. We are "top" when we are accountable (i.e., possess assigned responsibility) for assignments; "bottom" when we entirely depend on others and all we can do is help them (we are fixers); "middle" when we are intermediate between at least two others and, for example, resolve their conflicting needs, demands, and priorities. Members are "customers" when they need the output of the team's activities to advance. As a customer, we serve as the validator of the activities, while as intermediates in the middle position, we can play an integrating role. Team members play all these roles interchangeably and simultaneously, depending on what the circumstances call for. It is essential to realize that in each of these conditions, we, as accomplished self-leaders, make essential contributions to the team's success.

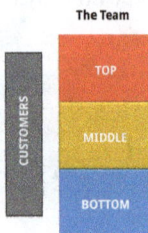

**Fig. 6.1:** In Oshry's system model for teams, the members continuously move in and out of conditions such as top, middle, bottom, and customer, irrespective of their formal function or position in the organization.

### Sports teams

*Team sports such as basketball, volleyball, and soccer essentially rely on the talent, skills, and self-leadership of the players. Although the coach is of great importance for tactics, strategy, deciding who plays, and all that, during the match, the team has to make all instant decisions and rely on an interesting combination of experience, routine patterns, genial hunches, and hard work to win a game. The self-propelling team relies on automatisms, occasional brilliant actions, and the determination to win whatever happens. The team has to win the match, while the coach stands on the sideline.*

## 6.2  Understanding the people you work with – we are all different

The decision to involve a student in a project or hire an applicant is a very important one. Having the wrong person on board can interrupt or, worse, paralyze, or even break up an entire team. How do you know beforehand if a candidate (e.g., an applicant for a new position in your group) will fit in? First impressions count firmly, especially when a new student or an applicant comes to see you about a position in your research group. We all will have an immediate impression of somebody we meet for the first time. And whether it is right or wrong, incomplete, or affected by the occasion (think, for example, about the possible stress of a candidate for a formal job interview or the talented actor who plays a role), a first impression determines our opinion about this person for some time, until we get to know him/her better. Experienced recruiters know this well and try to base hiring decisions on more objective criteria. Of course, the decision to proceed with the student/applicant is very important. In addition to the candidate's knowledge, experience, skills, and expertise, it is essential to judge if he/she will fit in your team and can interact positively with you and the other members of your group.

### 6.2.1  A little psychology: the "big five" personality traits

Without any claim of treating this subject correctly in any depth, we believe that a bit of insight into the psychology of the human character is helpful. While the classification feel–think–do–drive that we introduced in Chapter 2 has shown its value in classifying people's behavior in the workplace, psychology deals more broadly with people's mental and behavioral characteristics.

Historically, the ancient physician Hippocrates (460–370 BC) formulated a theory based on the notion that having too much or too little of four essential body fluids determines one's personality. It was later associated with the classic elements of earth, water, air, and fire. Although largely abandoned, one can still recognize its effect in older literature, opera, and drama, and it is interesting to know at least its origin; see a summary in Tab. 6.1 (Childs 2009; Eysenck 1967; Steiner 2008). Intriguingly, the typification is similar to the personal performance and management dimensions: air–thinker–content, fire–power of will–results, earth–doer–operations, water–feeler –relations.

Tab. 6.1: Hippocrates' theory of body fluids determining one's personality.

| Types | Body fluids | Elements | Characteristics | Comments |
|---|---|---|---|---|
| Sanguinistic | Blood | Air | Optimistic, active, social, creative | Notorious late comers; can be forgetful |
| Choleric | Yellow bile | Fire | Short tempered, fast, extrovert, egocentric, task oriented | Often have management potential |
| Melancholic | Black bile | Earth | Analytical, wise, introvert, conscientious, individualistic | Down-to-earth, objective attitude, often susceptible to negativism |
| Phlegmatic | Phlegm | Water | Calm, patient, thoughtful, tolerant | Innerly oriented and motivated |

Much more widely accepted nowadays than the obsolete body-fluids theory is the so-called big five or OCEAN model, which describes personalities not so much in terms of characteristics rather than in five traits (the first letters form the acronym OCEAN; Fig. 6.2):

- **Openness** to new experiences and ideas versus preference for the comfortable, safe, and well-known. Personalities scoring high on this trait are usually interested in modern art, literature, and traveling to new destinations, and are curious about new insights, latest fashion, and gadgets. At the same time, persons on the other end of the spectrum prefer to stick to their trusted routines and are not so keen on change. Of course, we hope to have scientific researchers in the first category, but please take into account the usefulness of team members who value and safeguard well-established routines for maintaining quality and safety standards and sound research practices in your laboratory.
- **Conscientiousness** means acting according to careful planning and being accurate without forgetting details versus improvisation and spontaneous action, with the risk that important aspects are overlooked. As a leader of a research group, we obviously wish to rely on our coworkers to do their research accurately and completely. Still, it will not be the first time that following a spontaneous idea for a crazy experiment eventually leads to unexpected and precious results and insights.
- **Extraversion**, being interaction and people-oriented, versus being introvert and internally oriented. Extraverted personalities are often also perceived as assertive, energetic, and full of initiative. However, we also know the stereotype of the eccentric scientist, who prefers to work alone, reads and reflects a lot, and can be very successful.
- **Agreeableness** means being friendly and compassionate with others and seeking compromises, versus competitiveness, self-orientedness, and the tendency to distrust others until the opposite is proven. Obviously, a team will be happy with some

"agreeable" types who make all others feel welcome and at their place. However, if the whole team is always "agreeable" and avoids conflict, no one will ever bite the bullet, and difficult issues may remain unresolved.

– **Neuroticism, or emotional instability**, implies sensitivity to stress, feelings of insecurity, and a tendency to amplify negative stimuli versus being relaxed, balanced, and realistic. High emotional stability reflects itself in rather calm personalities, who on the other hand might be perceived as unconcerned or uninspired. In contrast, on the other end of the spectrum, we find easily excited personalities, maybe seen as dynamic and perhaps even a bit unpredictable.

The principle that the five OCEAN traits form a basis for describing human character has been widely accepted, even across different cultures, although regional adaptations do exist. We think it is important to emphasize the value of having team members with healthy variation in their personality traits.

For scientific researchers, and in particular scientific leaders, we can envisage a few other important traits, for example, intelligence, creativity, endurance, and, not to forget, integrity. A theory called "trait leadership" (Zaccaro 2007) identifies no less than 17 assets of successful leaders, including the OCEAN big five, several typical management skills, as well as charisma, described as the ability to inspire their environment based on a compelling vision, that induces enthusiasm and commitment.

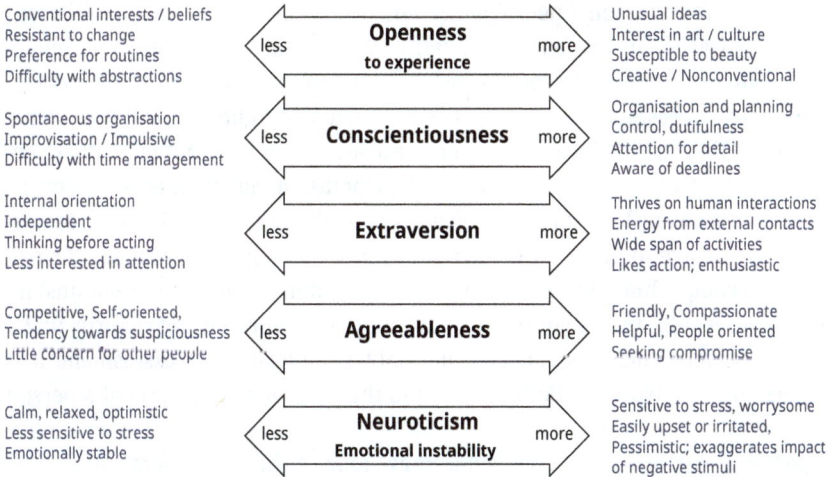

| Conventional interests / beliefs<br>Resistant to change<br>Preference for routines<br>Difficulty with abstractions | less **Openness** more<br>to experience | Unusual ideas<br>Interest in art / culture<br>Susceptible to beauty<br>Creative / Nonconventional |
|---|---|---|
| Spontaneous organisation<br>Improvisation / Impulsive<br>Difficulty with time management | less **Conscientiousness** more | Organisation and planning<br>Control, dutifulness<br>Attention for detail<br>Aware of deadlines |
| Internal orientation<br>Independent<br>Thinking before acting<br>Less interested in attention | less **Extraversion** more | Thrives on human interactions<br>Energy from external contacts<br>Wide span of activities<br>Likes action; enthusiastic |
| Competitive, Self-oriented,<br>Tendency towards suspiciousness<br>Little concern for other people | less **Agreeableness** more | Friendly, Compassionate<br>Helpful, People oriented<br>Seeking compromise |
| Calm, relaxed, optimistic<br>Less sensitive to stress<br>Emotionally stable | less **Neuroticism** more<br>**Emotional instability** | Sensitive to stress, worrysome<br>Easily upset or irritated,<br>Pessimistic; exaggerates impact<br>of negative stimuli |

**Fig. 6.2:** The five traits, often called the "big five," used in psychology to describe characters. Together, they form the acronym OCEAN.

### 6.2.2 The Myers-Briggs types

The Myers-Briggs typification system[1] distinguishes four dichotomies on which people act, take decisions, or approach situations, namely, **E**xtraversion–**I**ntroversion, **S**ensing–**IN**tuition, **T**hinking–**F**eeling, **J**udging–**P**erceiving, and then identifies which property is dominant in a person. This leads to 16 personality types, indicated with 4 letters, like ESTJ, INTP, and ENFJ. The system has found international acceptance and is often used in recruitment and selection procedures. The disadvantage is that it is almost a binary system; one is either a feeler or a thinker, and subtle mixtures of these characteristics are not possible in this classification, even apart from whether feeling and thinking should be considered opposites on the same scale.

### 6.2.3 Personal performance and self-leadership dimensions

Although not beyond all possible criticism either, we prefer to work with the simpler four personal operational dimensions introduced in Chapter 2, namely feel, think, do, and drive. They provide a simple but practically functional system for describing and obtaining insight into people's behavior in their work and, to some extent, their interaction with other people, as we explore in the following pages. We often refer to these four as management dimensions.

We use these four personal performance dimensions along with accomplishment mindsets in our (self-)management assessments. At one axis are the personal dimensions (doer, thinker, driver, and feeler), reflecting people's thrives and preferences. This is what most assessments do, and it is indeed valuable to understand how someone functions in their natural way. However, these assessments do not indicate if people are in control of what they do. For example, whether a person is an extrovert or an introvert does not reflect whether he/she is capable and successful. Our assessments, therefore, have another axis: the management dimensions or, more formal, performance accomplishment mindsets (operations, content, results, and relations) indicate whether people are in control of how they are expected to perform daily. Therefore the performance assessment is not only a self-leadership assessment. It is also a management instrument. Both oneself and the management can see if a person is in control.

The personal performance dimensions show four of the seven personal dimensions. To complete the personal dimensions, we add three personal (self)-leadership dimensions: the "*I*," the "*self*," and the "*impetus*":[2]

---

1 The Myers & Briggs Foundation. http://www.myersbriggs.org/
2 Impetus is defined as a positive, energetic inner drive that causes a process or an activity to be done, such as impetus for improvement, impetus for further study, etc. Impetus creates a highly energetic thrive.

- The "I" is externally oriented, individual, and tangible; is aware of what is going on and needed in the here and now. It is oriented directly at concrete outcomes, results or output. The "I" acts outside-in. What is the goal, and what exactly do we have to do to reach this goal? For all this, the "I" is aware that a person needs a tangible position, with power, status, respect, and recognition, and that a person will strive for the interest of himself/herself and the organization he/she belongs to.
- The "self" is internally oriented, intangible, and aware of what is needed there and then. The "self" lets things happen and works inside out. It is directed at processes and believes in the automatic outcomes, results, and output that these will deliver. For this, the "self" is aware that a person needs trust, belief, recognition, and a connection with the inner sources of know-how, wisdom, and intuition and that a human will strive for harmony and needs like being here, belonging to the collective, and being able to contribute.
- The "impetus" (sometimes also called "spirit") believes in connecting to meaningful purposes and sources of positivism and wisdom, thus creating enthusiasm and inspiration for everybody. It is strongly connected to one's thrives, ideals, and personal sources of inspiration.

### 6.2.3.1 Interaction between people based on operational personality dimensions

Scientists in the exact domains of mathematics, physics, chemistry, life sciences, and engineering are generally not interested in incorporating psychological considerations in their thinking and way of working. An open mind for how personalities may differ and an appreciation for what drives people help build productive collaborations. We have seen many examples of great and highly successful scientists who, over the years, have developed into real people managers who know how to inspire their team members to excel in many ways and often establish warm and long-lasting personal relations with them.

Figure 6.3 shows the profiles of two people with entirely different performance characteristics. The question is now: Can these two work effectively together? Well, at first sight, the profiles complement each other, so if the do–drive-oriented person is willing to carry out the ideas of the feeler-thinker, then this could be a good team, but not necessarily so.

Let's suppose you just had a progress meeting with your team. You ended with a list of tasks for the team in the following days and agreed on who does what.

The DOers in your team immediately start carrying out their tasks, e.g., ordering materials for the new project, making changes in the laboratory, and filing data from the previous project. Their working mode is to be active and do things as much as possible, and they can do so because there are clear protocols or routines for their activities.

The THINKER approaches the work package by first reflecting on what is needed and how it is best done, perhaps making a plan, and then starting to carry it out.

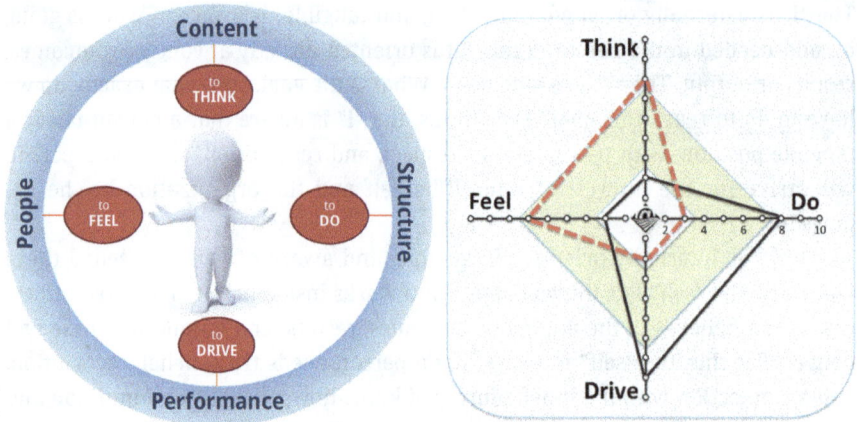

Fig. 6.3: The four operational personality dimensions and the profiles of two clearly different persons.

Throughout the day, he/she frequently stops to think: Am I doing the right thing? What are the alternatives? Should I do it differently? They think first and reflect a lot on what is happening.

The FEELER acts on sensitivity, subtle consciousness, intuition, and possibly experience. They feel what needs to be done, are willing to improvise, interact a lot with other people, and get energy out of these contacts. The FEELER may involve other people to help, and the work will be done because he/she has friends.

The driver, a typical achiever, has a strong will to get the job done and a strong intuition for what is needed to get results fast. They thrive on generating tangible results and ticking off items from the to-do and deliver list, and in this, they can sometimes be quite dominant toward other team members.

Successful teams have members who are driven by different dimensions. Suppose your group has only people whose first impulse is to DO things. Think of people who like to synthesize new materials, measure properties, collect literature, clean the laboratories, etc. What if they run off to do these things, but nobody THINKS carefully if these are the right things to do, or nobody has the intuition to FEEL that the activities could be more clever? Balanced teams will combine a few approaches and hopefully correct each other, and together, they will go in the right direction, whatever that is.

It is interesting to contemplate what happens if two personalities with their main operational personality dimensions interact. Suppose a DOer and a THINKer have to carry out a task together. The positive (i.e., synergetic) mode is that the THINKer comes up with a clever way to execute the task. At the same time, the DOer starts by trying out a few different approaches, after which, together, they decide on the best way, and they complete the job. In the negative mode, the THINKer makes it too complex, and the DOer loses interest in working together.

**Tab. 6.2:** Interaction between you and people with certain performance dimensions, seen from your perspective.

| | | The other | | | |
|---|---|---|---|---|---|
| | | **Do-er** | **Thinker** | **Driver** | **Feeler** |
| **You** | **Do-er** | Agreeing what to do & doing it | Realizing the good ideas | Executing the orders | Taking care for work |
| | | No reflection on what is done | Superficiality; lack of understanding | No feedback or too late | Forgetting to establish real personal contact |
| | **Thinker** | Developing and proposing feasible solutions | Agreeing on content and strategy | Providing alternatives; offering sense of control | Gaining trust through knowledge and integrity |
| | | Making it too complex | Too little awareness of realistic context | Insisting to be right, on having the final word | Focus on content without good personal contact |
| | **Driver** | Offering support for implementation | Acknowledging insight; offering to implement it | Swift and energetic action | Acting responsably and accountably |
| | | Superficial pragmatism | Ignoring content; overruling the other | Competition; unvalidated judgements | Dominance; perceived as unfair by the other |
| | **Feeler** | Taking care that tasks are carried out properly | Being interested in the other's vision | Utilizing his strive for dominance | Acting on a basis of mutual understanding |
| | | Losing reality in focus on execution | Not enough feedback on content; no output | Emphasis on pleasing the other | Missing practical relevance |

Effective Interaction      Ineffective Interaction

A typical ineffective interaction occurs if two DOers together run off and "start doing," without contemplating if what they do makes sense.

Interesting combinations are those where a DRIVER is involved. This personality type can be pretty dominant in making others work for them, as they are determined to get the job done and produce tangible results. When combined with a THINKer, effective modes of collaboration result when the latter comes up with the best and perhaps some alternative methods for completing the job, while giving the DRIVER the feeling that he/she is in control. If the outcome is positive, both benefit (and let's hope the essential contributions are acknowledged). However, ineffective interaction is also conceivable, for example, if the THINKER insists on one strategy and the DRIVER is not patient enough to try it out because they see a quicker way.

Another example is the following. You, a DRIVER for achievement, have a strong urge to finish a difficult project, and your coworker is a typical FEELer. Synergy will likely occur if the other accepts your lead and helps you while you value the other's role and make him/her feel appreciated. The ineffective interaction is not difficult to imagine: your dominance is perceived as lack of recognition for the other's contribution and is seen as unfair. Table 6.2 summarizes the constructive and destructive modes of interaction between team members with examples. It can be a useful exercise to construct such a table for your team members.

## 6.3 Successful team leaders have emotional intelligence

How can a leader ensure that the team thrives and that the members work in harmony and synergy? In Chapter 2, we briefly summarized the characteristics of successful leadership as providing a compelling vision shared with and by the entire team, having a highly developed sense of emotional intelligence being good at personal relationships (including mutual trust and respect), and disposing over ample managerial skills. We still have to explain emotional intelligence and how we recognize it in people.

Goleman (2000) has given a rather rigorous description of what emotional intelligence entails. He defines the concept as a complex of five ingredients: self-awareness, self-regulation, motivation, empathy, and social skills. We largely follow his treatment here and go through each of these qualities:

– Self-awareness: Ability to recognize and understand one's moods, emotions, drives, and their effect on others. Self-aware people are usually confident, have a realistic impression of themselves, and often display a self-deprecating sense of humor.
– Self-regulation: The ability to control one's own emotions, particularly if these can negatively impact others (anger, drift, dislike, etc.), and to suspend immediate judgments (think before acting). Self-regulators can often cope well with ambiguity and are open to change, or even to the unknown.
– Motivation: A passion for realizing a vision for ideological reasons rather than for salary or status and an inner drive to accomplish goals persistently and optimistically.
– Empathy: The ability to recognize and understand the feelings and emotions of others and to interact accordingly with them. Empathic leaders are aware of cultural aspects and sensitivities and often successfully develop and retain talent. Servant leadership is the style where empathy is most prominent as an essential quality.
– Social skills: Being good at establishing and managing relationships, building networks, and finding common ground among team members and/or stakeholders. As a leader this person knows how to persuade others, to induce change, and to build and lead high-performing teams.

Understanding emotional intelligence and its role is vital if your team is international. Highly developed emotional intelligence can help you understand and guide people with different cultural backgrounds, often with values that are different from yours.

Deutschendorff (2013) has given a crisp summary by stating that emotionally intelligent leaders are generally nondefensive and open to their environment. They are well aware of their own emotions and know how to control these, while they readily recognize the emotional state of others and respond appropriately. Furthermore, they are available for the people reporting to them, not necessarily by always having their office doors open, but by offering clear and nonforbidding procedures for quick contact when needed. Finally, and a clear hallmark for successful leadership, emotionally intelligent leaders can check and control their egos and allow their coworkers to shine.[3]

### 6.3.1 Emotional intelligence and emotional culture are linked

Chances are high that the leader's emotional intelligence sets the tone for the emotional culture in the group (Barsade & O'Neill 2016). Positive feelings shared by the group translate to better performance and higher quality of work, while chances of burnout and people quitting jobs prematurely are lower. When negative emotions dominate in the group, such as fear of making mistakes, anger about working conditions lacking attention from the management, or right-out suppression, the team's commitment will be lower. At the same time, some members may even become indifferent about reaching the leader's goals. In essence, such situations lead to behavior focused on not making mistakes, hiding talents, and suppressing creativity, involvement, and dedication.

Note that the cognitive culture can easily be described tangibly in terms of the shared intellectual values, norms, artifacts, and assumptions on which the team operates in daily life. However, the emotional culture is mainly transmitted through less tangible channels, such as body language and facial expressions. If the boss always looks worried or angry (whether on purpose or not), the impression may have an overriding effect on the atmosphere or emotional culture of the group.

The emotionally intelligent boss will be aware of such undesired effects. He/she will also understand the value of positive "micromoments" in daily life. A brief inquiry about someone's health after last week's flu, a compliment for something done well, a bit of small talk about last night's football match, a little gesture on somebody's birthday – it does not cost much, but it does wonders for the emotional climate. The same holds for the entire team, of course. Practicing good manners, like saying "please" and "thank you," is a simple enabler for people to work together, whether they like each other or not (Drucker 2008). Photos of the last group outing on the wall will also do better than a poster with dos and don'ts and a list of consequences when the rules are broken. Hence, the leader's and team's well-developed emotional intelligence will very likely translate to a positive and stimulating emotional culture in the workplace, where people feel confident enough to behave authentically.

---

3 A self-leadership assessment addressing emotional intelligence is available from www.scientifi cleaders.com.

## 6.4 The "3B" approach to authentic resources

In a team, mutual feelings of trust and respect, belonging to the group, and being valued for what one contributes are key factors for people to become successful team players. We have coined this "the 3B approach to free authentic resources," and it is based on the notion that feeling valued in relationships is often a key to successful performance and personal satisfaction. These "3Bs" are as follows:

1) **B**e oneself means that people feel so confident in and accepted by their environment that they behave fully authentically. There is no need to play a role, hide weaknesses, or pretend that one is better than one is. The comfortable feeling of being accepted in a group stands opposite to the fear of rejection, which leads to behavior focused on not making mistakes.

2) **B**elonging to the group is the opposite of feeling abandoned or locked out. Feeling fully accepted by the group will undoubtedly help one use one's authentic resources, as meant in the previous point. Team members will not have to fake their behavior and attitudes to be accepted.

3) **B**e valued, which is feeling recognized for one's contributions and confident that one's strengths are appreciated. At the opposite end of the spectrum, one would, for example, have the situation that a team member develops an inferiority complex due to the feeling of missing skills, or not knowing enough.

Scientific leaders, or any leader, should thus ensure that their students and coworkers experience being allowed to be who they are and form accepted members of the group. By guiding and challenging them to contribute in small but overseeable portions, they will learn step by step and simultaneously build on an authentic, stable personality. The fundament for authenticity is recognition, which is taking people seriously for who they are and what they think and want. Listening is an important skill, just as giving attention and constructive feedback. When successful, the approach leads to an atmosphere where cocreation through natural interaction can occur. Dialogues help the other to be clear on his/her intentions, feelings, and emotions. Discussions could help the other person determine how he/she would like to reach his/her intention.

## 6.5 Supervision – helping students to become colleagues

Graduate students (i.e., MSc and PhD students) form the core of academic research groups, thus making their proper supervision a substantial task of the academic research scientist. Many universities offer courses on supervision for their staff, and excellent literature is available (Wisker 2012).

According to Lee (2008), supervision of doctoral students encompasses the following ingredients and concepts:

– functionality – guiding students to manage their research projects;

-   enculturation – introducing students into the community of their scientific discipline;
-   critical thinking – stimulating students to critically question and analyze their work;
-   emancipation – encouraging students to question and develop themselves; and
-   development of quality relationships – where students are enthused, inspired, and cared for, also ensuring that they move along a path toward increasing independence.

For the authors, the ultimate success of supervision and guidance is if our students, new employees, or younger team members develop into colleagues whom we regard as equal on the subject of their activities and whose knowledge and/or expertise is actually superior to those of our own. How is this achieved? Several factors play an important role. Is your relation based on hierarchy, on power, or – opposed to this – on mutual trust between you and your protégé, in an atmosphere of openness and honesty and shared desire to do what is the best for the organization and for the two of you? The latter would be ideal, of course. It may take a while before a student recognizes this and realizes what your real expectations are. We all know the timid types who come with the attitude to learn mostly facts and procedures and obediently do what the professors tell them to do. If that is the dominant culture in your environment, one should not expect too many initiatives from the group. Stimulating them to be more assertive, confident, and, hopefully, become true self-leaders may require much effort from you. Again, trust and feelings to belong to the group and to be appreciated for what they contribute are key factors for many students to become successful team players.

Figure 6.4, adapted from the work of Blanchard et al. (1985), shows how to supervise new students or employees in different stages of their personal development. A new team member first needs direction, and instruction, as his/her competence level is probably low. When he/she gradually gains experience and confidence, the supervisor can adopt a more coaching and supportive role, until the team member feels well in place and becomes self-propelling. Note the continuously growing level of competence in Fig. 6.4. At the same time, commitment and confidence may vary, for example, a few months after the start when reality strikes and the new team member discovers that the life of a researcher can be challenging and different from what is expected. A few successes help to reestablish confidence and commitment. Supervisors should be aware that this situation often occurs.

When knowledge and technical competencies are low, your coworkers need you to set clear goals, plan what to do, and set priorities. They need training to become skilled and monitoring to check progress or to prevent going astray, and they need regular feedback. However, to build commitment, your coworkers need your support: someone who listens, who not only corrects mistakes but also praises and encourages, motivates why their work is essential, and shares information and insight to make them feel connected to your goals.

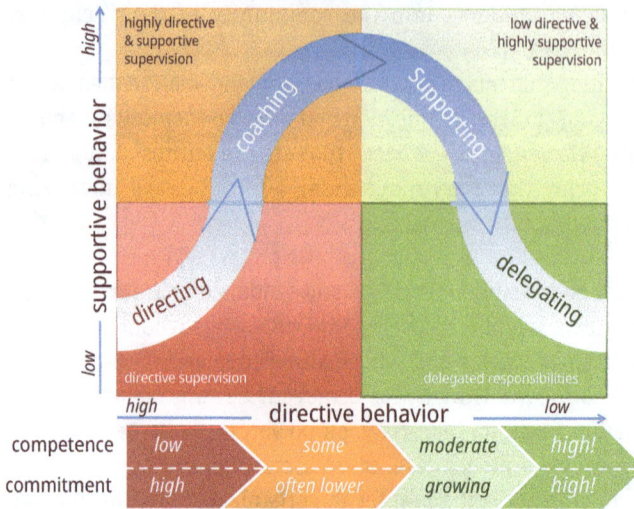

**Fig. 6.4:** Trajectory of becoming acquainted with tasks, for example, for a student entering the research phase or a new project team member. The novice first needs direction, but gradually the supervision style can change to coaching and supporting, while tasks can be delegated after some time. All the time the student's level of competence increases, but the supervisor should be aware that commitment may vary in stages when the job presents difficulties or is simply different than expected (adapted from Blanchard, Zigarmi, and Zigarmi).

## 6.6 Concluding remarks

Successful teamwork is all about synergetic interactions between people who acknowledge the value that they can achieve more together, even if their eventual goals are different. Understanding people and being able to recognize their thrives and emotions are essential. For this, emotional intelligence is a must-have leadership skill. Awareness of people's rather elementary needs concerning their functioning in groups is necessary for all team leaders. After all, leaders have the same need to thoroughly enjoy their role in leading the group. We firmly believe that the "3Bs" formulated (being oneself, belonging to the group, and being valued for one's potential contributions) are important, but underestimated, factors in the success of teams. We see them as essential prerequisites for establishing trust, mutual respect, an open culture for critical and constructive exchange of ideas, and pride in achieving shared goals.

# References

Barsade, S., O'Neill, O.A.: Manage your emotional culture. Harv Bus Rev; 2016; 94; 58–66.

Blanchard, K., Zigarmi, P., Zigarmi, D.: Leadership and the One Minute Manager; William Morrow, New York, 1985.

Childs, G.: Understand Your Temperament; Rudolf Steiner Press, New York, 2009.

Deutschendorf, H.: Five ways to spot an emotionally intelligent leader, June 27, 2013. www.tnlt.com.

Drucker, P.F.: Managing Oneself; Harvard Business Press, Boston, 2008.

Eysenck, H.J.; The Biological Basis of Personality; Thomas, Springfield (Illinois, USA), 1967; p: 35, 39.

Goleman, D.: Leadership that gets results. Harv Bus Rev; 2000; 78; 78–90.

Lencioni, P.: The Five Dysfunctions of a Team: A Leadership Fable; Wiley, New York, 2002.

Lee, A.: How are doctoral students supervised? Concepts of doctoral research supervision. Stud Higher Educ; 2008; 33; 267–281.

Oshry, B.: Seeing Systems: Unlocking the Mysteries of Organizational Life. 2nd Edition; Berrett-Koehler, San Francisco, 2007.

Rostrup-Nielsen, J.R.: 50 years in catalysis. Lessons learned. Catal Today; 2016; 272; 2–5.

Steiner, R.: The Four Temperaments; Rudolf Steiner Press, London, 2008.

Wisker, G.: The Good Supervisor, Supervising Postgraduate and Undergraduate Research for Doctoral Theses and Dissertations; Palgrave MacMillan, Basingstoke, Hampshire, UK, 2012.

Zaccaro, S.J.: Trait-based perspectives of leadership. Am Psychol; 2007; 62; 6–16.

# Guest column: Research should be managed by motivation – not by control

Jens Rostrup-Nielsen †

*Dr. Rostrup-Nielsen was research director of Haldor Topsøe A/S in Denmark and a founding member of the European Research Council. In this guest column of 2017, he gave his view on how to manage responsive research in an industrial environment.*

Haldor Topsøe (1913–2013), founder of the Danish catalysis company that bears his name, has always believed in a multiple approach to industrial research. It was his firm belief that understanding the science behind the development work is of crucial importance. First, this gives a strong basis for coping with problems, and it has great value for marketing activities too when potential customers notice that a company has deep and comprehensive understanding of the technology they offer. Much of the basic research can be published, which motivates the scientists and forms a basis for valuable collaborations and joint publications with university groups. Such collaborations can be sustainable only when they are two-way processes, where both sides benefit. Hence, it is essential that a company is active in basic research as well.

A second strategic component in Haldor Topsøe's philosophy was to be always at the forefront with core competencies, from advanced computational modeling to the newest methods for materials research, e.g., state-of-the-art electron microscopy and the use of big research infrastructure at synchrotrons. Being active in the further development of key research methodology yields competitive advantages for doing high-profile research. It makes the company an attractive partner for leading academic groups, and it creates the ability to catch opportunities for new directions in contact with the best of the field.

The third ingredient in the multiple approach was exploration aimed at "radical innovation," working on solutions for the next-generation technologies. To some extent, this can be done in-house or by outsourcing test attempts in the academic world and of course by actively monitoring what competitors are up to.

Managing all these parallel activities can be a challenge. In some way, handling the worlds of logically planned R&D and creative basic research is balancing order and chaos. I sometimes compare the management of innovative research with gardening. The meticulously laid-out baroque garden looks beautiful, but there are no surprises. A garden without any control, on the other hand, turns into an interesting wilderness with lots of activity, but probably not in the direction you want. Obviously, the truth is in between.

For the leader, the art is to manage with a loose line and to ensure that each scientist has room for one's own initiative. One can institutionalize this, for example, by

---

†Deceased December 3, 2022.

using the 65–25–10 approach (which by the way I picked up from a research director at Shell): about 65% of the resources in a project are used for development work with well-defined milestones and budgets, and 25% is used by the team for its own solutions to the problem. The final 10% is available for any activity that a researcher finds useful within the scope of the company, sometimes referred to as the "Friday Afternoon Experiment." In this mode, the scientists have a lot of opportunities to contribute to their own insights and have their own initiatives. Strict planning of activities is inevitable for large development projects, but it is a killer for projects in the explorative phase. Research should be managed by motivation, not by control.

Hence, for me, managing responsive research is striking the balance between "focus and hocus pocus," avoiding that innovation processes become too bureaucratic and ensuring that each scientist has room for one's own initiatives.

## Reference

A longer version appeared as Rostrup-Nielsen, J.R.: 50 years in catalysis. Lessons learned. Catal Today; 2016; 272; 2–5.

# 7 On the road to scientific (self)-leadership

## 7.1 Introduction: leadership based on authenticity and agility

The path to leadership goes via self-leadership. Having experienced the journey from apprenticeship to what it takes to lead and manage your own research, your own research unit, your own laboratory, or even your entire department provide an invaluable basis of know-how for helping others grow in their first career steps. But how do you become a true self-leader and a leader for others? We think authenticity, the ability to act naturally and in line with your personality, is a crucial prerequisite. The next phase is to become fully aware of what you are passionate about and how it can help you set goals and realize them. Finally, one needs to build up experience. This requires enhancing the appropriate skills and deepening your insight into people and processes.

The concept of authenticity is prominent in this work on management and leadership theories (Gardner et al. 2011). Many theories on organization place strong emphasis on transparency and authenticity. People feel better when they can behave authentically and they expect to be allowed to do so in their work. Customers generally appreciate it when they are served in authentic ways. Students like to be treated, guided, and taught by professors and lecturers who are themselves and do not act out of some survival mechanism or hide their authentic personalities behind a copied role model in or outside the organization.

> **3B–6T–9E**
>
> In the world of management and leadership theories, it is not uncommon to coin concepts as the "ABC of consultancy" or "1-2-3 to Success," often to the dismay of the erudite scientist. Forgive us for dubbing our approach to (self)-leadership as "3B-6T-9E concept:"
> - 3B as the three requirements for authentic behavior (be oneself, belong, and be valued)
> - 6T as the six stages in the trajectory to realizing a grand plan (Triggers, Talents, Thrives, Thrills, Trails, and Track)
> - 9E as the nine elements for developing skills and building up experience (including skills for autonomous and team performance and devising strategies for creating high impact)

In addition, it is often said that "change" has become a constant factor in modern society. Organizational structures in which we operate tend to change rapidly nowadays. This could be because of external factors like legislation, unexpected changes in external funding (be it positive or negative), or it could have internal causes like new appointments at the executive level (a new head of department, a new dean, a new director) or a change in the way budgets are distributed internally. Rapid and continuous change forces organizations (the department, the research group) to respond fast with "agile" strategies, to benefit from new opportunities, or to minimize adverse effects. Those who are slow to respond are often disadvantaged, whether organizational units, stakeholder groups such as teachers and students, or even individual scientists:

https://doi.org/10.1515/9783111325644-007

*The well prepared, who constantly anticipate policy changes, and always have a plan ready, benefit most when the opportunity arises, while slow followers will find that the budget has already been spent when they come with their plans.*

Agility, the capability to anticipate change and respond fast in an optimal way, is much easier to realize for people who behave authentically (spontaneously), certainly when they also have a clear vision of where they want to go (see also the section on dynamic strategy, realizing intentions in an ever-changing context). Rapidly changing conditions are handled best by leaders who dispose over agility, for which authenticity is an almost indispensable prerequisite. There is no longer any time for programming and deprogramming people in modified organizational schemes with new guidelines prescribing how to behave.

### 7.1.1 You and your (self)-leadership personality

Like everybody else, you have an inherent, natural blueprint of your four personal performance dimensions (do, think, drive/willpower, and feel). These are connected to your personality, expressed in terms of your autonomous "I" (the individual accountability), your inner "Self" (norms, values, inner feelings of responsibility), and a third quality, which we indicate by the term "Impetus" (your inner drives, passions, sources of inspiration that make you thrive). The first four, do, think, drive, and feel, largely determine your potential to perform, while the more personality-based "I," "Self," and "Impetus" relate to your personality and leadership capabilities. Table 7.1 summarizes the concept.

**Tab. 7.1:** Personality as expressed in three personal leadership dimensions.

| "I" | "Self" | "Impetus" |
|---|---|---|
| – Individual accountability for results | – Responsibility for proper processes | – Inner drives connected to external goals |
| – Norms | – Value-driven, ethical | – Passion |
| – Directed at tangible achievements | – Inner awareness | – Inner source of energy and inspiration |
| – Willing to compromise | – Conscience | |
| – Egocentric side of our personality | – Wisdom to act | – Self-enforcing, self-propelling |
| – Safeguarding one's own position and interests | – Social side of our personality | |
| | – Directed at harmony and fulfillment | |
| – Having | – Being | – Thriving |
| Lives in the "here and now" | Time perspective less relevant | May change slowly over time |

The "I" lives in the here and now and has a good intuition about the context and what it takes to be effective. The "I" looks at feasibility and is willing to settle for realistic accomplishments, for example, being satisfied when a task is completed at an 80% level because it suffices for the purpose whereas our "Self" might find it much more rewarding to do better and go for 100%. The "Self" reflects your values and responsibilities. As these are mainly invariant over time, the "Self" lives in the continuity, one can also say in the "there and then." The "Self" is directed at harmony, sustainability, and wholeness. The "Impetus" is more difficult to describe but could be considered your inner sources of energy, enthusiasm, and passions. For many, it relates to their beliefs or philosophy of life as the primary source of inspiration. In essence, it relates to your authentic and free drives.

## 7.1.2 Self-leadership rests on tangible and intangible skills and capabilities

Self-leadership is both about *you* and about *what you want to achieve*. Too often, we find these two worlds apart: Managers in powerful positions usually focus on performance and achieving tangible results (e.g., the drivers of Chapter 2), while trainers and coaches – "people managers" – pay much more attention to the individual who has to be ethical and authentic based on intangible skills and personality traits only.

As scientists or engineers, we have been trained to make rational and objective decisions using our cortex, the brain's upper, analytical, and reflective part. However, many decisions in daily life are based on emotion and intuition. Personality and spur-of-the-moment are important factors, and many people make decisions based on what they emotionally want and what feels good to them. This is a fact of life, whether we like it or not.

The art is integrating both the hard and tangible with the soft and intangible side because both sides need each other and can factually empower each other. Leaders who manage to incorporate both the tangible and the intangible approaches in a balanced way are much more likely to reach success with their teams, that is, realize the goals of the organization and the individual members. Think, for example, about the professor who, in his/her role as mentor and inspirer of outstanding research, completes a challenging research project. At the same time, the very motivated PhD students graduate with excellent theses and publications and the postdocs position themselves favorably for the next phase in their careers. Everybody wins; moreover, the university is happy, science is enriched with new knowledge and insights, and society benefits from new talent and perhaps even new opportunities for innovation and exploitation.

Self-leadership also enormously concerns your capabilities: job, life, and personal skills. Some skills come naturally to specific people; but in general, they derive from a mix of talent, experience, and participation in specific training and courses. We are all different – some capabilities come naturally to us while others can be tough to acquire, if at all.

It is essential to realize that people learn and acquire expertise in circular ways. Similar issues happen repeatedly, and experience helps in dealing better with them each subsequent time. However, in the negative mode, recurring challenges may also become a continuously growing barrier and eventually a blockade to performance, for example, if people avoid a challenging or confronting solution or always defer the problematic issue to others. The favorable look at circular learning is to see it as an upward spiral, which over time widens the scope along with the enhancement of personal capabilities and personality and a broader reach (Fig. 7.1).

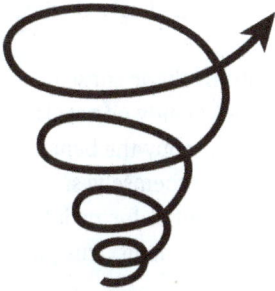

**Fig. 7.1:** Gaining experience is like an upward spiral – it grows and widens its scope with time.

**Enhancing** servant leadership (ESL) and enhancing self-leadership
ESL (Gardner et al. 2011) is a theory based on thorough practical research into the essentials of self-control, combining the personal dimensions that energize people with the performance and impact dimensions that lead to success. We prefer the word "enhancing" to "developing." The latter means "unfolding" (but what is left of an onion when you peel off all the rings?), while enhancing expresses enrichment. For the authors, it has factually more than one meaning. Enhancing implies enlarging your possibilities to grow, to perform, and to create impact because you combine an increasingly broader scope of experience- and insight-based sensitivity and sensibility with focused decisions and attention. It also implies consciously following up on positive feelings and rising above negative emotions, making the approach focused on getting the best out of oneself, while guiding and supporting the students and other team members working with you. Finally, it refers to enlightenment, to upraise oneself and behave and act more consciously.

Enhancing *Servant* leadership is focused on stimulating people to get the most out of themselves. It does not mean that the leader is the servant of his people. They are much more experienced and have the leading role in letting their people grow while they and the organization perform. The leader is accountable and knows how to stay in control of their organization. However, they do not resort to commandeering or suppressing people; he/she leads their people to self-control with responsibilities and accountability within the organizational settings.

Felderhof, J.P.K.: *Enhancing servant leadership through inspiring, enabling and focusing enhancing self-leadership*; Informance Publishers, Eindhoven, 2007.

## 7.2 Part 1: the power of authenticity and feeling recognized – the 3Bs

As we said before, authenticity plays an essential role in leadership theories. What exactly is it? The adjective "authentic" means genuine, honest, neither false nor copied, representing one's true nature or beliefs, and "authenticity" is the quality of being authentic. Kernis (2003) describes authentic behavior as "acting by one's values, preferences, and needs as opposed to acting merely to please others or to attain rewards or avoid punishment through acting 'falsely.' Authenticity is not a compulsive effort to display one's true self, but is the free and natural expression of core feelings, motives and inclinations."

People who behave authentically are as they are. They intuitively or consciously know their strengths and accept/admit their weaknesses, whether openly or not. They are themselves under all circumstances. They do not play a role or copy the behavior of others to make themselves appear better than they are – they are themselves.

What is needed to be authentic? Three aspects in life are essential for people to function authentically and all three regard mutual respect, fair treatment, trust, and recognition:
1) the feeling of being allowed to be oneself (in contrast to feeling rejected);
2) the feeling of being connected to the people who are essential in one's functioning (opposite to feeling left alone or deliberately excluded); and
3) the recognition for making a (potentially) valuable contribution (with the feeling of an inferiority complex on the opposite end of the scale).

Note the word "feeling" in all three points, emphasizing that the subjective perception of the individual (student, postdoc, employee, or yourself) matters and not how the boss or the environment assesses the situation.

**3 B's for free authentic resources**
Be oneself – Belong to the group – Be valued.

People who are recognized by their environment extract positive energy from this feeling and encouragement to go on and do even better. Of course, recognition needs to be deserved, e.g., showing genuine interest in the project and the people, working hard, being proactive, etc. Fear of rejection, on the other hand, can undoubtedly serve as a motivator for working and trying hard, but at the risk of unwanted side effects: people experiencing negative stress, trying to prove themselves, being bossy, or withdrawing within themselves and shying away from confrontation, to mention a few defensive reactions. People with a fear of being abandoned might start to counteract by avoiding conflict (e.g., refraining from critical discussion or from contributing new ideas) or even by flattering and pleasing others. Such behavior is far from authentic. Nevertheless, we have seen many organizations, including academic groups, where the culture is fear-driven, where people are afraid to make or admit mistakes, scared to take wrong decisions, or propose correct choices that are not favorably looked upon by the management or frowned upon by the environment. Such factors work as imposed and assumed constraints (Chapter 2).

To avoid misunderstanding by the reader, by emphasizing the value of recognition, we do not mean that leaders should only praise their team or that individual members constantly give each other compliments. Such "mutual admiration" societies can undoubtedly belong to the culture of organizations, but they are hardly productive since complex issues and disputes are deliberately avoided to keep the atmosphere pleasant. Biting the bullet is sometimes needed, and conflicts and critical assessment of ideas, handled respectfully, are essential for progress. Weak points in a research project and flaws in the execution of experiments or the interpretation of data could better be addressed immediately by the team than they appear later and cause damage and embarrassment, e.g., during a discussion at a conference, by a critical reviewer who recommends rejection of a manuscript, or – even worse – rejection of the next funding proposal. Openly challenging opinions and critically interrogating procedures and interpretations are all feasible and, in fact, essential in a culture where people are respectfully acknowledged for their efforts and valuable contributions.

Hence, recognition is a crucial enabler for feeling at ease and, thus, for authenticity, and the combination of recognition and constructive feedback (which should include criticism and points for improvement as well) is essential for high-performing teams and is particularly important for people in early stages of their career. Figure 7.2 illustrates this. A novice (e.g., student and new employee) depends mainly on external recognition, until they have enough experience and self-confidence to assess the situation and their own performance realistically. The final stage on the career development scale is the reputed professional, who receives recognition from peers. Each field in science and engineering has such stars that seem to attract funding and bright students like magnets. They have become a "brand" in their discipline.

**Stages in Career Development**

| Dependent on External Recognition | Recognition by oneself<br>building confidence | Professional Recognition<br>(becoming a "Brand") |
| --- | --- | --- |

Parallel from life:
*Childhood*    *Becoming Mature*    *Parenthood*    *Mentorship*

Fig. 7.2: Comparison of career development and stages in life. At the first transition (dashed line), we see dependence disappear and maturity grow. In contrast, the dashed line on the right indicates the transition to becoming an established professional as recognized by peers. Their name, their organization, or their entire university may become a brand in the community.

## 7.3 Part 2: a personal plan toward a successful scientific career and (self)-leadership – the 6T approach

To be fully effective in building and shaping your laboratory, department, institute, company, or whatever it is that you want to accomplish, your personal performance dimensions, thinking, feeling, driving, and doing, should come together in drawing up and realizing a plan, like in Fig. 7.3. These are the steps:

– You need *to think* about your future and translate it into a vision and a mission.
– It would be best if you were inspired and became enthusiastic: you need to *feel* the sources that give you energy and motivation.
– You want to *feel* how trusted colleagues, students, and others in your environment respond to your ideas. Are they genuinely interested, skeptical, or enthusiastic? Do they believe in it?
– To make the plan work, you need the inner *drive*, the power of will, the determination and resilience to set and realize goals.
– Finally, you need *to do* what is needed not only the exciting try outs of novel experiments in the laboratory but also the less attractive routine activities required to keep the organization going.

> **Operational dimensions in teaching**
> For an educator, understanding how the student learns (feeling) is just as important to understanding the topic (*thinking*) that somebody teaches. Compensating empathy (*feeling*) with overemphasizing the knowledge (*thinking*) and persuading the student to work hard (*driving*) may have adverse effects.

This may sound obvious, but why is it sometimes so hard to succeed in practice? If we disregard external circumstances for the moment, one needs all four personal operational dimensions, at least to some extent. People who have brilliant ideas and great empathy to enthuse others but need more inclination to **do** things with their hands or miss the inner **drive** to push through will probably not get very far. Or imagine the overambitious, perhaps somewhat egocentric drivers who expect everybody to help them unquestioningly because they work so hard for what they see as the good cause. Great thinkers may not be able to get to the execution phase and so on. In conclusion, one needs to have the four management dimensions of thinking, feeling, driving, and doing in some balance.

What if we lack one or more of these qualities? The danger is not so much to dispose over too little of one or two personal dimensions; the real problem arises if a person does not recognize this or deliberately ignores shortcomings in certain aspects of management. Admitting weaknesses needs courage and self-confidence. It may well be that the environment or culture does not stimulate the person to be open and reveal vulnerabilities. A reaction observed often is that people hide their weaknesses and compensate for deficient dimensions with a quality they are good at, in this way creating a "survivor identity."

Fig. 7.3: Balancing the four management dimensions of thinking, feeling, doing, and driving is an important asset in realizing successful projects.

## 7.3.1  The risk of ignoring essential dimensions or declaring them irrelevant

Another complication with managers who are weak in one of the four management dimensions and – consciously or unconsciously – found a way to cope with it is that, over time, they gradually start to believe that the lacking dimension is unnecessary.

Drucker (2008) presented this – in the meanwhile classic – example of the planner, who naturally relies on his greatly developed "think" dimension:

> Like so many brilliant people, he (the planner) believes that ideas move mountains. But bulldozers move mountains; ideas show where the bulldozers should go to work.

Another example of this sort is scientists who compensate for poor decision-making (i.e., they are missing the drive dimension) with overwhelmingly warm feelings toward students and colleagues (the feel dimension) or developing theories of everything (the think dimension). These scientists may be popular in their direct environment but could have serious difficulty securing funding or delivering output. The opposite is the hardliner, who merely relies on the do and drive dimensions and never invests in good personal relationships.

What actually happens is that people who miss out on one or more of the four personal performance dimensions create "safe havens" for themselves where they feel at ease. In reality, they miss opportunities by purposely blocking productive relations with coworkers who communicate preferentially on their missing dimensions.

The solution for compensating missing qualities is obviously to collaborate. An acclaimed full professor who is weak in, e.g., the feel dimension can be greatly helped by a warm and people-oriented secretary or assistant, provided both understand each other well and mutually acknowledge their roles. Willingness to listen to each other is an essential prerequisite for such alliances' success. Another example in the academic setting are the visionary professors who creatively work on the fundamentals of their field of interest and are great motivators and mentors for their environment, but are not particularly good at finishing reports before the deadline, pushing students to finish up, or submitting proposals in time. If such a leader has an assistant professor in the group who focuses on these important aspects of academic life (i.e., in the do and drive dimensions), then the whole group can run very successfully. Again, a condition is that both the leader and the trusted right hand mutually acknowledge their roles, communicate well, and are willing to play these parts.

To realize strategies, achieve success, and create impact with results, all four dimensions must be present in the unit's leadership, be it in one person (e.g., the full professor and the dean) or in the collaborative efforts of the team. In the latter case, the roles should be clear, which means that the players' strengths and weaknesses are known and mutually acknowledged, and communication is open and frequent.

### 7.3.2 The 6T trajectory to authentic leadership, high performance, and strong impact and becoming passionate about it

In our courses for personal development and scientific leadership, we find it essential that participants discover what makes them passionate about their work or, if this is not what they feel, what would have to change to become passionate. The word "passion" has many meanings, but we define it here as the very strong liking of and dedication to an idea, a concept, or an activity. Being passionate is not a state of mind that one switches on; it grows with being exposed to the idea or being involved in the activity, and eventually, this passion becomes firmly anchored in one's inner "Self" and the "Impetus" of Tab. 7.1.

We usually start the training session by explaining some of the principles as in Chapters 1, 2, and 5, including what a vision, mission, strategy, SWOT analysis, and integral plan are. Next, we carry out a workshop where participants make a strategic plan for themselves, their research group, or even their department. We assume that they have implicitly or explicitly a vision or at least some ideas on what they would like to achieve. We then start by showing the participants Fig. 7.4, a scheme that we have named the 6T Trajectory, meant to integrate personal plans with the organizations. We invite the attendees to reflect little on their current performance as a person and as a group and to be fully aware of how things are running in their environment, whether or not they are satisfied with the situation or think that there is room for improvement in whatever personal or organizational aspect of the work, including its perspective for the future. Perhaps, participants are already familiar with the SWOT inventory and analysis or an assessment based on the comprehensive scheme (Fig. 5.4 in Chapter 5), or otherwise, participants rely on their impressions for the moment.

We then ask all participants to take a step back and to think about what *triggers* them, what aspects of the work make them enthusiastic, what gives them warm feelings, or in general, what sources provide energy. Such *triggers* probably reflect the person's (hidden) *talents* and *thrives.* In this respect, Freud noted that it is hard for

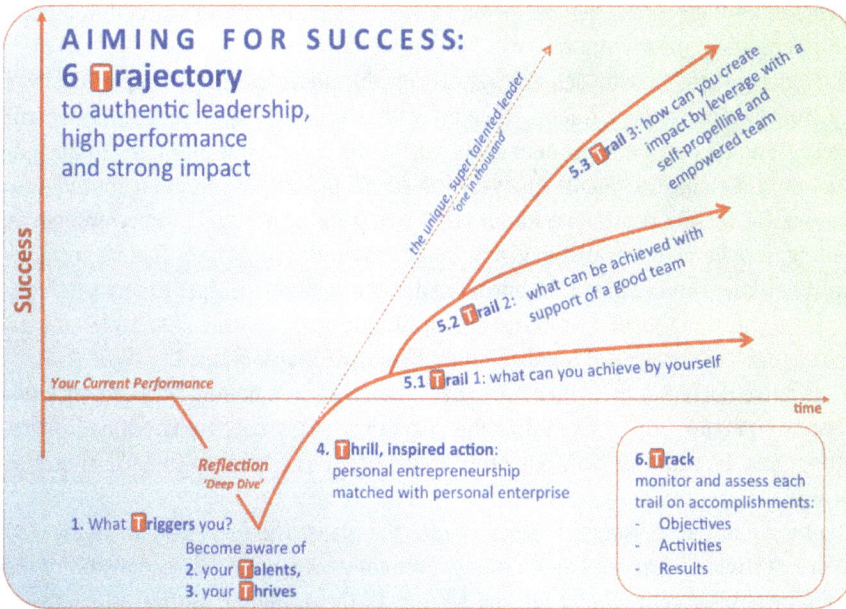

Fig. 7.4: The 6T trajectory to authentic leadership and high performance and impact.

people to go directly to the subconscious mind, but events, actualities, and situations often trigger subconsciousness. In other words, we sometimes surprise ourselves with hidden talents and tacit knowledge that come forward when dealing with practical situations. Once we are aware of our triggers, talents, and thrives, the next challenge is how to make optimal use of them and make them operational.

In an interview with a Dutch newspaper, Nobel Prize winner Ben Feringa was asked for his most important advice to young scientists. We translated as follows:

"Follow your dream and don't make it too easy for yourself. It is important that you discover for yourself what gives you energy. Once you know this, aim to go to the limits of your abilities. Don't be distracted and follow your intuition. This can take you very far in science."

Prof. Ben Feringa; interview by J. Jansen, Het Parool, June 20, 2017.

Now we come to the fourth and fifth T's: one's daily *thrill* and one's *trails*. To give these a stimulus and direction, one must first try to define the intention that a person, department, or organization wants to fulfill. Knowing talents and especially thrives, they can examine their dreams: what would one like to accomplish – in life, or, perhaps smaller, in one's working life, or, yet even smaller, in the actual job. Such an approach works for a department or an organization as well. Try to relate these dreams to talents and thrives and find out what is realistic to accomplish in the short- and long term. Then transform the dream into an intention that is realistic and feels good. This intention should then define one's daily thrill and provide direction for the trails for achieving this intention.

Thus, the fourth T in the scheme stands for **the thrill, daily feeling of** positive energy from realizing one's talents, thrives, and internal energy sources. This thrill can be regarded as an entrepreneurial spirit, inspired by what one tries to accomplish. Identifying triggers, talents, thrives, and thrills brings out the inner motivation, the reasons for feeling passionate about what one is doing or trying to accomplish. A compelling, maybe even idealistic, vision and mission are important factors in motivation. When one can connect it to one's talents and thrives – in fact to one's personality – in a way that these strengths are optimally utilized, one disposes over an immensely powerful source of inspiration and energy for accomplishing goals.

If the drive derives only from external sources, such as a demanding boss or environment, the person who must realize the goals may easily become frustrated or develop burnout. In addition, without inner drive, it will be hard to inspire others to cooperate.

The next phase is to sketch **trails** to realize the plans. The first (trail **5.1** in Fig. 7.4) is to work out what a person can do alone, autonomously, with full responsibility for execution and full accountability for end results. In the academic setting, scientists on this trail will probably be helped by students and perhaps a postdoc. Developing skills, learning by doing, and gaining expertise will all be essential parts of the journey.

On the way, when the organizational unit (e.g., the research group) acquires resources, opportunities arise for trail **5.2**, namely, moving in the direction of a self-propelling team. In the academic setting, one may think of the associate professor who just received tenure. The group has several PhD students, postdocs, perhaps a technician, and especially the more senior members and the technician each contribute their strengths as valuable assets for the entire team. This type of group could evolve in a self-propelling unit, operating like a flock of geese, where each member takes the lead on specific aspects of the work, as circumstances demand. This would be another example of flock or shared leadership (see Fig. 7.5).

Some people join flocks naturally, but this is not generally the case. Before being ready to co-operate in flocks, one should first learn how to handle challenges and solve issues autonomously. The idea of a flock is not primarily to let others handle the difficulties, but to reach farther destinations together than one could alone – all members with their strengths and responsibilities. People generally like to cooperate with others who are autonomous and have something unique to contribute. Irritations about members who merely joined to benefit lead to conflicts, and negative energy, that distract from the objectives. Once people are conscious about their strengths and weaknesses, and the culture allows them to be authentic, without egos and fears blocking relationships, they can start to build on each other's strengths, talents, and natural thrives. Then co-operation in the natural way of geese flying together in flocks is possible, with the benefit of reaching further with less effort.

Fig. 7.5: Flock leadership in a self-propelling group.

### 7.3.2.1 Strategic partnerships for enhanced impact

Trail 5.3 is yet another step up in potential opportunities. We have called it the strategic partnership, where teams from the same or entirely different organizations collaborate strategically to benefit from joint expertise. The situation is best explained with examples. Imagine yourself in the research group of Professor A from University X, with unique strengths in synthesizing materials for a particular application, e.g., microelectronics or molecular nanoengines. At University Y, a group under Professor B has valuable expertise in computational predictions of materials properties. The two professors know each other's capabilities and decide on a strategic collaboration, where the computational expertise in Prof B's group is applied to predict materials with improved properties for the applications ventured in Prof A's group. Together, they achieve results that none of the groups separately could have obtained, and together with their PhD students and postdocs, they publish high-impact works in top journals.

The other example, again from the scientific research front, is that of an academic group with significant expertise in specific techniques or theories, which collaborates strategically with the R&D department of an industry. Together, they focus on fundamental issues underlying real-life problems from industrial practice. This knowledge enables the company to improve processes based on fundamental insight. At the same time, the academic group works on truly relevant problems and benefits from industrial funding and the highly relevant publications they publish in collaboration. This is synergy with high impact in action. The authors can cite several successful strategic collaborations of this type. In fact, they have been part of several such ventures themselves (van de Loosdrecht et al. 2016). These examples reflect the third trail (Trail **5.3**) of the scheme.

To travel these trails over the years, one may plan the strategy regarding milestones. Goals for these milestones can be split into smaller stepping stones, each with its own concrete objectives and activities. Progress can be monitored by keeping track of activities and results. Overarching, one can strategically keep track of progress with respect to the milestones by evaluating whether the original goals are still relevant or that readjustment of the strategy is needed.

### 7.3.3 Dynamic strategies

To realize a mission, a shared vision and a well-structured strategy are needed. Otherwise, creating success would be a matter of pure chance, which may happen occasionally (like winning a lottery). Hence, a consciously intended strategy is essential, but

Fig. 7.6: Rapidly changing conditions urge us to adjust our strategy constantly. Our initially defined goal becomes a point on the horizon, and we do our best to end up somewhere near. External factors cause us or give unexpected opportunities to readjust the strategy. The quality of self-leadership lends us the courage and self-confidence to decide to change direction, while our sense of entrepreneurship guides us in the right direction. Self-management enables us to do everything needed to keep the project on track in the new direction, until the next point of strategy revision arises. A well-articulated vision helps to keep the project overall in the desired direction.

external factors and unforeseen circumstances may render the carefully thought-out strategy useless. Therefore, strategies should be adjustable. Mintzberg et al. (1996) already introduced the concept of emerging strategies. Figure 7.6 illustrates the dynamic nature of an agile strategy that is adjusted when circumstances dictate.

## 7.4  Part 3: nine elements for enhancing your capabilities – 9Es

Third-generation governance and control incorporate insight and knowledge of all seven personal performance dimensions (do, think, drive, feel, "I", "self," and impetus). Combined with all other sources that you have available, such as sensing, intuition, acquired experience and wisdom, creativity, and relational qualities to fully benefit from personal contacts with others, be it in formal networks or daily at work.[1]

Everybody will have their natural preferences and inclinations. The question is to what extent you have developed these qualities and which opportunities and experiences you have had so far to serve as a nurturing ground to discover, exploit, and enhance these seven personal dimensions to their full potential. Learning to recognize these opportunities for enhancement is an essential step toward growth. Perhaps, the situation in which you were raised, studied, and acquired your experiences in life has led to a minor or possibly more substantial unbalance, in which specific personal dimensions have been more strongly developed than others. For example, people who grew up under challenging circumstances might have developed their "I" in combination with their "do" and "drive" dimensions simply because they had to fight for their existence in a hostile environment, while others had all the chance to utilize their talents in music or sports fully and developed strong "Impetus" and "feel" dimensions. These well-developed dimensions can undoubtedly serve as strengths in your future career. On the other hand, dimensions that received less attention could hamper your chances to become truly successful in the future or, in other words, the chances to stay ahead of the competition.

Natural control has its roots in the seven personal dimensions we described above: all of them can be empowered and energized, stay at the present level of development, or even worse, become frustrated, leading to blockades and dissatisfaction on your achievements. Hence, ideally, your natural dimensions should have the chance to mature toward optimal effectiveness in your daily activities (=self-management), in achieving your personal goals (self-leadership), and eventually, in becoming an inspiring mentor for others (leadership) (see Fig. 7.7). We propose the scheme shown in Tab. 7.2, entitled "9E – nine elements to enhance upon," as a long-term guideline for enhancing your capabilities and yourself.

---

1 Tests for scoring yourself on these dimensions with respect to personality, leadership, and entrepreneurship are available, see www.scientificleaders.com.

**Tab. 7.2:** Nine elements for personal enhancement.

| Nine elements 9E | Characteristics and subjects to enhance on* | | |
|---|---|---|---|
| | **Perform autonomously** | **Perform with a team** | **Perform with teams in partnerships** |
| Full self-control | **1.1 Manage yourself** Deliver on a personal basis; invest in management skills; master self-management (based on self-management assessments). | **2.1 Lead yourself and others** Master self-leadership and start to lead others (based on (self)-leadership styles assessment); articulate your vision and develop a strategy to carry it out. | **3.1 Lead and mentor others** Grow toward undisputed, authentic leadership at multiple levels of your organization or community. |
| Scientific entrepreneurship (= taking new initiatives) | **1.2 Playing field** Play your role in the organization; others will call upon you, develop yourself as a loyal and valued team player; explore, and get to know the playing field in which you operate. | **2.2 Contact management** Invest step by step in contacts with your peers from the wider community and start collaborations (be selective and ensure mutual benefit for all partners). | **3.2 Intentional leadership** Convert dreams to reality in your own organizational unit (research group, department, etc.) and outside (scientific community, society in larger context, etc.; based on the personal intentional leadership assessment). |
| Scientific innovation | **1.3 Initiate wider context** Take initiatives to bring the organization (the research group of your full professor, the department) further; take the lead in new initiatives of the department. | **2.3 Collaborate with impact** Collaborate effectively in and outside the organization; build international reputation, have comprehensive strategy for realizing your goals, learn how to build a relationship, if desirable, with different personalities than you are. | **3.3 Smart Impact Strategy** Create innovative impact by, e.g., strategic alliances, spin offs, new schools, new departments; international branches. Apply the integrated performance and impact innovation approach. |
| *Examples roles* | *Assistant professor, junior research scientist* | *Associate professor, senior researcher* | *Full professor, institute director* |

e-Learning courses and assessments on several of these elements and coaching programs are available via www.scientificleaders.com.

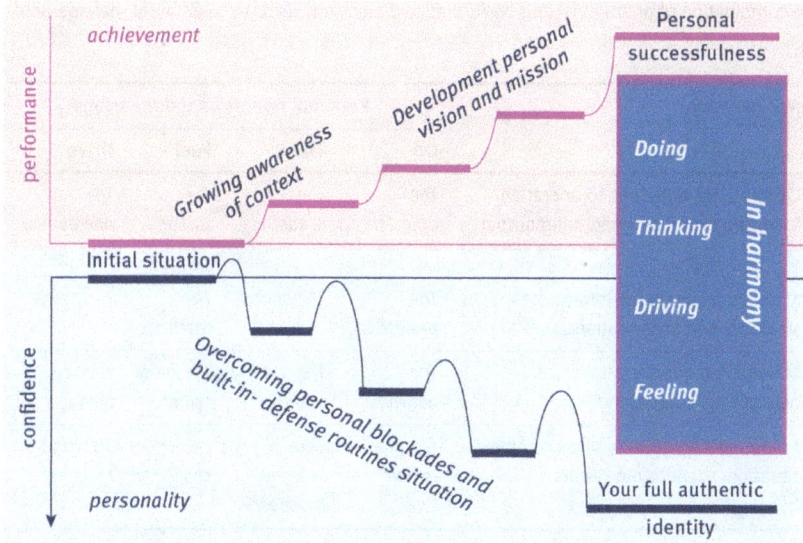

**Fig. 7.7:** Our success increases gradually while we gain experience and overcome personal blockades. Eventually, we use our personal performance dimensions in a natural and balanced way that works best for us as we are confident enough to behave authentically.

## 7.4.1 Toward complete self-control: the performance aspect

The upper row in Tab. 7.2 is labeled "self-control." Being in control over yourself is decisive for your success and career, and therefore, it is also decisive for the success of the organization in which you are working. It is important to develop and constantly enhance your self-management roles and to learn several typical management skills such as managing time and projects, handling conflicts, or becoming a master in communication. Some of these skills we have dealt with in this book, but specialized courses are usually offered by universities, and many companies organize such courses for their employees.

Self-management combines personal dimensions (do, think, feel, and drive) with performance dimensions (structure, content, people, and results). The operational dimensions represent the sources of energy you use in your work, while the performance dimensions provide a focus for mastering your daily professional activities and achievements (see Tab. 7.3).

The combination of personal and performance dimensions yields a 4 × 4 matrix with 16 distinct management roles. We will not analyze these in this book but mention that in an advanced course program, assessments are available to investigate how candidates score on these roles and how they can improve on their potential. Such an assessment shows one's capability level for each performance dimension.

**Tab. 7.3:** The combination of operational and performance dimensions leads to 16 different management roles.

| Performance dimension | | Personal operational dimensions | | | |
|---|---|---|---|---|---|
| | | **Do** | **Think** | **Feel** | **Drive** |
| **Structure** | Carrying out well-defined operations, following protocols generating output | The executer | The organizer | The caretaker | The operational manager |
| **Content** | Knowledge and know-how (professional, strategic, and organizational) | The professional | The analyst | The teacher | The pioneer |
| **People** | Maintaining good relationships, networking, supervising | The networker | The creative dreamer | The social center | The people manager |
| **Results** | Accomplishing goals in often complex situations, generating results | The go-getter | The reflective manager | The challenger | The crisis manager |

## 7.4.2 Toward full control: the self-leadership aspects

For the more experienced researcher, who already earned tenure at the university or is, for example, on the way to senior scientist in a research institute, self-control involves self-leadership as the fundament for balanced, successful, and endurable careers and organizations. Self-leadership is needed for being the director of your own life. In terms of the seven dimensions, you need the "I" and the "self" in balance.

> **Compromises pose challenges**
> By nature, scientists score high in the content dimension. We often see, however, that the step of turning knowledge into tangible results can be difficult. In scientific research, this may well be linked to the desire to deliver complete projects, without loose ends or unanswered questions. However, reaching such a level of completeness would take an unrealistically long time, and hence, often, one is forced to settle for less than ideal and be satisfied with, for example, 90% instead of 100%. Making such compromises can be difficult and handling uncertainty is something else than dealing with analytics. It requires determination, emotional strength, and power of will, which is built upon the "I" dimension and practical experience.

Your "I" and your "self" form a duality. One (the "I") is delighted if 85% is reached and the job can be considered done. The other (the "self") strives toward perfection, toward 100%. What to choose? You need them both if you want to be successful. Your "I" takes care of your interests and position. Your "self" takes care of relations, honesty, and continuity. So, choosing means you gain and lose one, but you need them both.

The duality of the "I" and the "self" can cause significant challenges in life. Nevertheless, they are like two sides of a coin; one needs both, otherwise, it is not a coin.

### 7.4.3 The duality paradox between the "I" and the "self"

People who feel insecure for whatever reason may have difficulty behaving truly authentically. They are likely to develop a mode of behavior that allows them to cope with the situation, for example, by being overly assertive up to an inclination to control others or, on the contrary, rather giving in and surrendering than seeking a confrontation or even choosing to please others to avoid unpleasant situations. In such cases, the "I-Self" duality becomes a paradox that is hard to solve and a source of stress, imbalance, selfish behavior, or undesired alienation and disruption. People may not feel recognized, belong to the group, and appropriately valued for their input, talents, and efforts. Such situations give rise to survival mechanisms and artificial – that is, compulsive – behavior.

Suppose our survival identities and defense routines take control over the way we behave in our relations with others. In that case, we are essentially controlled by negative emotions, like fear, anger, sadness, jealousy, or revenge. Sometimes, people have developed such defense mechanisms to survive in earlier life, childhood, high school, or previous jobs. People might base their personality on survival routines, which become a self-identification. Suppose you have such members in your team. In that case, it may be quite a challenge to recognize such self-created identities, and it may be an equally difficult challenge and a long process to help them realize that behaving according to one's authentic identity gives much more fulfillment.

This is where leaders can be tremendously important in helping their team members develop into mature, authentic, and inwardly free human beings: people who are accountable, responsible, and enjoying life while following and achieving relevant goals through meaningful activities.

When people are caught in the duality paradox, they escape confrontations or ambiguous situations in several ways, e.g., by wanting to control (trying to be the boss), by choosing to obey (be a pleaser, or even an enslaved person), or by avoiding power games entirely, and focus exclusively on content. These four surviving identities are unbalanced but not entirely out of control. The unbalanced identities can nevertheless form a productive system. The overcontrolling identity takes the boss role and assumes accountability for the social and productive system based on the unbalanced personalities, thus protecting others incapable of functioning outside their comfort zone in conflicting and ambiguous situations. Being protected from insecure situations, the pleaser can play the social center, the thinker takes care of the content, and the doer can produce predictable output. The production system is complete. The "boss" handles negotiations, investments, funding, etc., and is accountable for the profits and losses, while his subordinates may rather ask for a budget and a salary, than for taking any risks. Directed by accountable management, such unbalanced situations can yield productive teams: the boss takes strict control over the others and provides them the protection they need for their "reality-evading" roles. In this way everything can work relatively well in practice, although no one involved will be authentically satisfied with the situation. Together they create a social system that can be very comfortable if everybody is respectful toward each other's surviving identity. Even in practice, too comfortable, without taking risks, the dependent identities get what they want socially, on content and compensation. Also, the surviving boss identity gets satisfied with the feeling of power, control, and social status and likely also financial rewards. However, socially the situation may become unpleasant, when the accountable becomes a selfish, nasty boss, or when dependent identities overrate their contribution and demand unrealistic compensation. Feelings of anger, sadness, hate, or even revenge may build up, with unpredictable consequences if these are triggered. In such cases, the system may collapse.

For example, think about the university department where the chairs strictly run their own research group and only take care of their part of the curriculum. The professors compete for resources by negotiating with the dean on a bilateral basis, while the monthly faculty meeting has become an arena where the professors fight for their own interests. Collaboration between groups inside the department hardly exists. Such a department can operate successfully, at least for some time. However, we are convinced that a department with chairs who dare to apply the type of leadership that stimulates authenticity and collaboration at all levels will have much better chances of success, in education and in doing innovative research.

When people are not caught in a duality paradox, they can cooperate openly, freely, transparently, and with trust building on each other's authentic power. The duality paradox reveals itself not only within people when caught in a survival mode and operating in defensive modes but also in environments that have turned into a political arena, where open communication is avoided and showing vulnerability is regarded as a weakness that is up for exploitation.

### 7.4.3.1 Does the organization encourage authentic behavior?

An organization's culture may or may not encourage its people to behave authentically; this creates situations between four extremes, as shown in Tab. 7.4.

Situation 1 is the ideal case for organizations that encourage authentic behavior and for the people who exhibit it. The organization relies on the self-control and initiative of all people involved. This implies individual and collective consciousness, a trusted context, self-confidence, and the courage to be vulnerable. People who commit to the vision and strategy of the organization rely on their self-management and self-leadership capabilities and take accountability and responsibility where possible. Organizations operating in this mode represent the most potent governance and control situation. New team members coming from an environment or culture where authentic behavior is not encouraged (situations 2 and 4 in Tab. 7.4) may need clarification and time and guidance to adapt.

Authentic leadership leads to balance and harmony for people within themselves, with each other, and their environment. People feel recognized, happy, and valued. They will also be dedicated to the organization and its goals, when these are congruent with their personal goals.

**Tab. 7.4:** Organizational culture and authenticity.

| Organization | Encouraging environment | Discouraging environment |
|---|---|---|
| **People:**<br>**Authentic** | **1**<br>Organization can be run naturally based on authentic leadership; people can thrive on their and each other's talents, drives, and capabilities and fully realize their potential (*Flock leadership situation: authentic, free, autonomous, responsible, and natural cooperation*). | **2**<br>People conscious of the situation decide if they accept it and find their own sweet way in the system and/or attempt to influence the system (in small steps) to become more open to initiative.<br>(*Hide or fight: "If you don't like the system – change it"*) |
| **Not**<br>**authentic** | **3**<br>Opportunity for the leadership to create an environment of trust and mutual respect where people feel at ease and eventually develop authentic behavior. Leadership should be attentive and prevent that persons in the team abuse freedom or manipulate others to their own advantage (ensure a *safe opportunity for natural growth*). | **4**<br>Organizations and people in this mode are essentially in an unstable form of governance control; the environment suppresses creativity and initiative; energies are (partially) blocked; cooperation is hampered.<br>(*Lasts as long as complying yields material and/or psychological advantages for the employees.*) |

The other extreme, situation 4, represents an unhealthy and intrinsically unstable form of control, where all involved accept the situation. Control is based on power in a hierarchical and sometimes bureaucratic structure with explicit and implicit rules. People focus on their tasks and avoid the risk of making mistakes rather than taking new initiatives. Perhaps they realize the situation and operate carefully or they become dominant and political when they want to have power and put their own interest first. In many cases, however, people might be unconscious, identify with their survival mechanism, and experience the situation as normal.

**Survival identities.** People, who are unconscious of their survival identity, have prevailing values, norms, habits, and attitudes that keep their surviving disposition alive and cover up their authentic personality. This feels normal and right for themselves, and they likely feel attacked when someone questions or criticizes their behavior. In their survival mode, they might well react defensively rather than be grateful for the opportunity to take a freeing look at themselves. Establishing a culture of trust in your environment is probably the best way to help a surviving identity leave his/her island of comfort and safety. Start by indirectly providing insights that are not too confronting. Gradually one can become more direct with feedback and suggestions, up to a decisive point where the surviving identity hopefully decides to (re)assume authentic behavior.

Unfortunately, many organizations, also in academia, reside to at least some extent in situation 4, characterized by a far-from optimal governance and control system, with a political arena of power interfering with people's wishes to apply their natural sources of talents, wisdom, and drives to generate useful outcome. It prevents people from thriving on the basis of capability and dedication to the successes they hope to achieve. The principle of governing is primarily to control people on what they should do or accomplish. However, in a natural process, factors such as control, professional know-how, creativity, and accomplishments are not separated. In situation 1, on the other hand, the governance principle is to give people freedom to perform and grow, each in their own way, and to lead them toward a common goal. The way to get there is cooperation, interaction, and intercreation. In nature, this is called flock leadership.

### 7.4.4 Being in control of your organization

For you as a leader, the way to control the organization in a natural way is to give your people the proper context to perform and grow through own experience. The extent to which self-management and self-leadership dimensions are developed in persons determines how far they are enhanced in this respect. When the degree of enhancement of a person in these dimensions is by the requirements of the function, the function and person are likely to make a good match while you and your organization have an excellent chance to be in control over what you want to achieve.

Figures 7.8 and 7.9 show how both the organization and its employees can be in authentic control while allowing freedom and giving responsibility. The entry level of a new employee (or PhD student) may concern performing prescribed tasks or routine procedures appropriately (see also Fig. 6.4 in Chapter 6). Giving this person too much responsibility would not work, and both the organization and the employee would find themselves in the upper left part of the diagram. The opposite happens when somebody with well-developed self-leadership capabilities is asked to work on routine tasks. Such a person would lose interest in the job, while the organization does not benefit from the employee's potential – the bottom right of the diagram.

The ideal situation arises when employees operate at or near the diagonal of the figure, that is, in the green zone of the diagram. People and organization are both in control, and the employees have exciting and satisfying jobs, in which they are appropriately challenged.

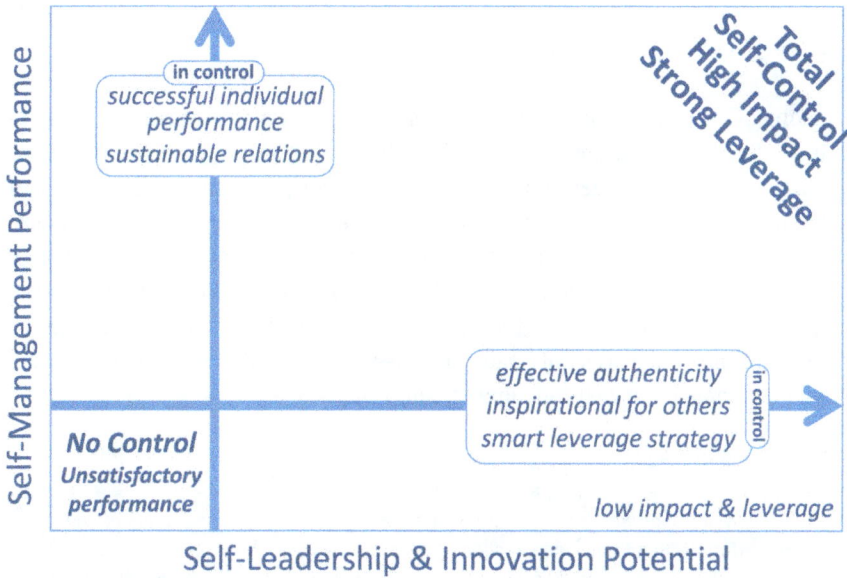

**Fig. 7.8:** To experience high impact and leverage, agile organizations rely on the initiative and commitment of their members, implying that both self-management and self-leadership capabilities are strongly developed.

**Fig. 7.9:** Relation between governance and control in an organization, the level of personal development, and responsibility of the employees.

## 7.5 Flocks, teams, and flock leadership: leading on authenticity without losing control

Geese, among other birds, fly in V-shaped flocks to save energy (Lissaman and Shollen-berger 1970). In flock formation, each bird flies slightly higher than the one before, to reduce wind resistance and get an uplift. The geese take turns leading the flock and fall back when tired. Estimates are that they can fly up to 70% farther in large flocks than alone. They must beat their wings less frequently and have lower heartbeats than birds flying alone. The V-formation is also believed to be beneficial for communication and keeping the flock coordinated. Professional cyclists riding in groups use similar principles to minimize the effect of head and crosswinds. The V-shaped flock thus represents an ideal model of a well-functioning, self-propelling team.

> **Flock leadership in the self-organized jazz band***
> An excellent example of a self-propelling team that illustrates the principle of self- and flock leadership is a jazz band, playing tunes with a lot of improvisation. Ranging from a trio to a band of any size, most of them perform without a conductor, although one of the musicians often has the lead. They usually start with a theme, e.g., a well-known song with a fixed score for every musician. After the first verse, the musicians take turns improvising on the theme. They lead during their solos, while the rest stick to their own part. Here, we see the self-propelling team of self-leaders in action, operating almost like a flock of geese in nature.
>
> *Thanks to Dr. Jens Rostrup-Nielsen, who often referred to jazz bands as perfect examples of well-performing teams; see page 128 for his views on leadership from an industrial research laboratory perspective.

By now, it will probably be evident that the authors support the flock leadership model as the preferred way to lead a team, small or large, and even an organization. Flock leadership rests on a natural, authentic style and on relying optimally on the team members' capabilities, personalities, and initiatives. Hierarchy is in place but does not dominate the culture. Power, if the word should be used at all, derives not necessarily from hierarchical position in the organization, but much more from strengths, knowledge, experience, insight, or valuable relationships with external parties. In this sense, flock principles match very well with Oshry's total system power that we discussed in Chapter 6 (to avoid misunderstanding, note that "power" stands here for "power to accomplish" and not "power over people").

### 7.5.1 How to lead on authenticity

Trust is the primary facilitator. Trust is vital for a healthy culture, with people being allowed to be themselves, which implies daring to admit weaknesses and acknowledge vulnerability. One can be vulnerable to weakness and behave emotionally dependent on others. Still, one can also be "powerfully vulnerable" in the sense that weaknesses are acknowledged while being confident and in self-control when organizing help or cooperation. In a leadership position, such people will control the situation. A culture of trust in the team is needed for people to feel assured that what happens is for the benefit of the team and its members. When people try to manipulate others in the group for their own interests or position, trust is undermined, cooperation suffers, and the team may degrade to an arena where people play games to win and gain for themselves. In such an atmosphere, people who strive for honest cooperation can easily be abused for their knowledge, creativity, and hard work. Also, openly admitting vulnerability would be considered a weakness that allows others to abuse it. Vulnerability would then imply losing position. Hence, one should not let vulnerability be abused in arena cultures. However, in leading for high performance and impact, fair play and building self-confidence are essential, and warranting a safe and trusted culture must be a top priority of the formal leader.

## 7.5.2 Staying in control of a self-propelling organization

Directing on authenticity is done by giving your people space for their own decisions and freedom for their own ways of working and solutions to ensure optimal and mutual benefit from talents. We see three opportunities for exerting control without limiting your team members' freedom to move:
- The first opportunity for control is the definition of the playing field. If clearly defined and regularly updated, directions and boundaries are set for activities and decisions.
- The second is to secure trust and challenge so that people can learn and strive for performance, growth, and impact. As discussed earlier in this chapter, the 3 Bs allow the buildup of a stable organization instead of one based on survival personalities.
- The third opportunity for control is to match the enhancement level of your employee's personality with the requirements for the position. The self-management, self-leadership styles, and personal entrepreneurship assessments available in the online course program not only give insights and direction in this regard but they also allow governance, control, and compliance. This is done through function assessments on the required personality level of enhancement, which can be used as a norm and then matched with self-and/or 360-assessments. The self-assessments give insight into the potential and self-awareness of people, and the 360 assessments provide a view of others and what people show in practice.

The ideal of self-propelling organizations functioning in flock-leadership mode is realistic. We have seen many good examples in our professional careers, and we are grateful to be and to have been part of such organizations. One must, however, be aware that the concept of self-propelling teams in flock mode still depends on the presence of a care-taking leader who is and remains actively involved in the activities of the team and who is constantly safeguarding the overall direction as laid out in the vision, mission, and the dynamic and therefore regularly adjusted strategy. We have also seen situations in which seamlessly functioning teams full of synergy fell back to a collection of individual projects executed by isolated people who only communicated with their functional superior or even to strictly hierarchical structures, as indicated in Fig. 7.10. Such undesirable transitions can be caused by incidents, acute conflicts in the team, lack of resources/funding due to a failed grant application, a change in leadership at a higher level in the organization, a reorganization of the organization under which the team resorts, and a multitude of other possible causes such as plain miscommunication due to people not understanding each other's minds/way of thinking. Well-running organizations are a precious commodity. Good leaders know this; they will do anything to anticipate risks, minimize their impact, and maintain their organization, team, department, and institute in optimal shape.

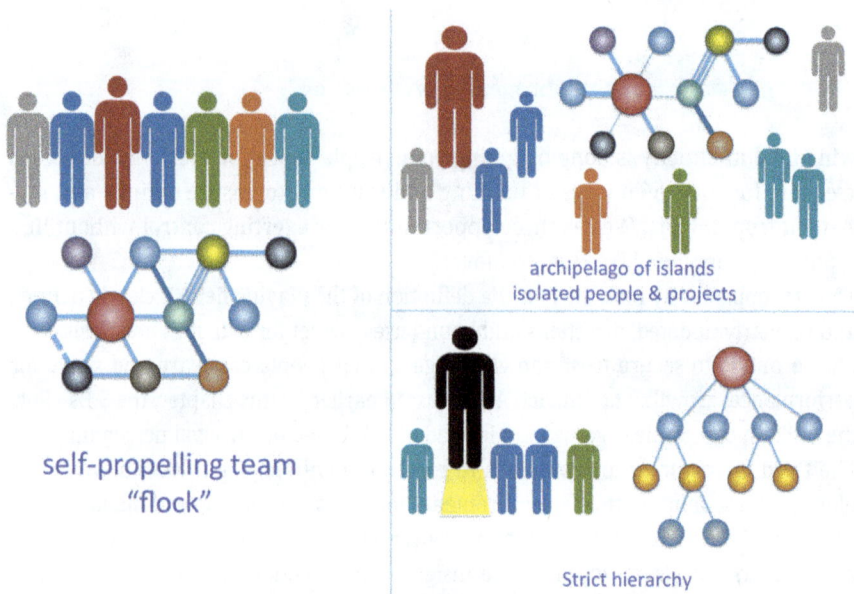

archipelago of islands
isolated people & projects

self-propelling team
"flock"

Strict hierarchy

**Fig. 7.10:** Self-propelling teams in flock mode can, over time, decay to other modes of operation such as in the satellite system without mutual connections other than to the leader or a strictly hierarchical model.

## References

Drucker, P.F.: Managing Oneself; Harvard Business Press, Boston, 2008.
Gardner, W.L., Cogliser, C.C., Davis, K.M., Dickens, M.P.: Authentic leadership: a review of the literature and research agenda. Leadership Q; 2011; 22(6); 1120–1145.

Kernis, M.H.: Toward a conceptualization of optimal self-esteem. Psychol Inq; 2003; 14; 1–26.

Lissaman, P.B.S., Shollenberger, C.A.: Formation flight of birds. Science; 1970; 168; 1003–1005.

Mintzberg, H., J. Lampel, J.W., Quinn, J.B., Ghoshal, S.: The Strategy Process, Concepts, Contexts, Cases; Prentice Hall Europe, London, 1996.

van de Loosdrecht, J., Ciobîcă, I.M., Gibson, P., Govender, N.S., Moodley, D.J., Saib, A.M., Weststrate, C.J., Niemantsverdriet, J.W.: Providing fundamental and applied insights into Fischer–Tropsch catalysis: Sasol–Eindhoven University of Technology Collaboration. ACS Catal; 2016; 6; 3840–3855.

# Guest column: Fundamental research on social investment and its impact on policy

Anton Hemerijck

*Prof. Dr. Hemerijck is a professor and program director at the European University Institute in Florence, Italy.*

In my career as a social scientist, from the humble academic beginnings as an assistant professor to the more prestigious positions as director of the Dutch Scientific Council for Government Policy in the Hague, as a dean and a deputy-rector at the Free University of Amsterdam, and now at EUI in Florence, I have always tried to lead by example, making sure that I continued to do research, write, and publish while taking on greater management responsibilities. As a senior scholar working with ambitious young scholars, I find it imperative to be tuned in to the latest theoretical and methodological debates and to have a comprehensive overview of the field. I never fell for doing science as an ivory-tower art for art's sake. Any social scientist, from economics to cultural anthropology, should at least aspire to foster social returns on cutting-edge research to make science matter for society at large. I would even argue that engaging with the ongoing challenges of socio-economic restructuring makes one a better scientist in one's own right, as changing environments inspire and feed curiosity, which is the most important forte of any academic. While this idea is simple, putting it into practice for the teams, departments, and organizations I have been responsible for is hard work. It requires constant reminding that good science, and by implication science leadership, begins with a curious, independent mind. When scholars are not independent and inquisitive, they often seek (emotional or psychological) safety in methodology and theory. I always tell my PhD researchers to never fall in love with theory and methods but to learn to love and cherish their explanations. There is no contradiction between fundamental and applied or engaged research. Theories and methods are tools for understanding the world and social change, but only explanations that can stand the test of time define scientific progress and are key to social progress.

## A case for effective yet affordable social policies

Let me share my recent experience as an engaged scholar as an example of putting this philosophy into practice. Europe's aging societies worry policymakers seeking to balance social equity and fiscal sustainability. In an interdisciplinary research project financed by the European Research Council, we could demonstrate that European welfare states are indeed expensive but also effective and affordable.

Together with a highly diverse team of postdocs and PhD students with macroeconomic expertise, microsociological statistical competences, and qualitative comparative policy process-tracing skills, we were successful in communicating and explaining to policymakers that an inclusive but active welfare state with a strong social investment commitment can be prosperous and competitive, and capable of containing poverty, while it can also swiftly bounce back from economic shocks, without harming fiscal balance over the business cycle. The shorthand research question of the project was what kind of welfare state is fair and sustainable for knowledge economies and aging societies in the post-industrial age?

Reasoning from a life-course perspective and a conception of the welfare state, we argue that affordable and high-quality childcare reinforces labor market attachment for young couples with children. Active labor market policies, lifelong learning, and public health policies enable adult and older workers to pursue longer careers. Social investment policy provisions bolster poverty mitigation by easing risky life-course transitions, precisely by fostering labor market inclusion.

## Unexpected impact on policymakers

An interview by *The Economist*[2] (6 March 2021) for the leading article on the surprise comeback of the welfare state in the wake of the Covid-19 pandemic sparked the interest of EU politicians in our project. Suddenly and unexpectedly, we had a direct impact on work at the highest level of European policymaking concerning the subject of economic returns on social investments.

The most important resource that our team has been able to offer in terms of policy outreach is human capital, particularly its diverse skill set. Given that the measuring rod of success is *scientific* rather than *policy* impact, we were initially reluctant to trade off research capital for policy engagement. However, after having been consulted at the highest levels of the EU, we decided to broaden our impact strategy. While not wishing to appear overconfident, we believe that the significant attention we have received from international policymakers and our engagement with think tanks conjures up an instance of real-time social impact, all while the actual research project was ongoing. Our strengths are twofold. On the one hand, there is a heterogeneous mix of skills, theoretical perspectives, and methodological approaches that we have been able to muster and develop further in our work. Second, building on diverse methods and approaches, we have developed a common language of life-course relevant investments in stocks, flows, and buffers in our search for new evidence for a sustainable – inclusive and ac-

---

2 *The Economist* is a weekly news magazine from the UK that is read worldwide. Its articles cover current affairs, international business, politics, technology, and culture.

tive – welfare state facing major socioeconomic restructuring. This represents a language that is currently gaining traction in policy circles.

The example demonstrates that reaching out from the "Ivory Tower of pure science" pays off by raising outside policy awareness and name recognition and by improving the quality of our primary research. In this respect, there is no trade-off between academic rigor and policy relevance, just as there is no contradiction between efficiency and equity.

# 8 Implementation of self-leadership and the 3B-6T-9E enhancement philosophy in organizations

## 8.1 At the level of teams

Sharing a mission, vision, and culture invites and stimulates your team members to adopt a culture of trust and mutual respect and enables them to build self-confidence. The individual goals may vary among the team as long as these fit within the leader's strategy. An example is the win-win situation of a PhD student graduating with an excellent thesis and some high-impact publications, forming an essential building block in the overall project. Teams become particularly productive when they adopt flock-like behavior. For this to happen, team members must experience – perhaps unconsciously – that the 3B principle operates: be oneself, belong to the group, and be valued. To achieve this, we believe that three conditions should be fulfilled:

- First, convince the group members that they are allowed to be as they are without having to prove themselves. This prevents people from suffering from fear of being rejected or of not being taken seriously "because they know so little." It may take some time before newcomers feel secure in this respect, and one should be aware of strong cultural differences that may play a role with international group members.
- Second, give your people the feeling that they belong to the group and have a place there. This means that they will not be excluded from important meetings or social gatherings and that their questions on things they do not know or understand yet are always taken seriously. Newcomers are part of the group because they believe in the common goals and are willing to contribute to them. The fear of being abandoned or excluded can be strong for novices.
- Third, give your people the opportunity to grow in their own way, with enough room for their own initiative, encourage them, and show recognition for their ideas, efforts, and contributions. In such an atmosphere, delivering critical and constructive feedback is easier. Indeed, in scientific research and development, people need freedom to move and challenges to grow through their own experiences. This implies that sometimes, the more experienced team members should hold back immediate reactions and ignore the occasional (less-critical) mistake to give the novice the chance to discover it independently and learn. When allowable, this takes away the pressure of being looked over the shoulder continuously. Also and again, if circumstances permit allowing people some frustration of failure has an educative effect and makes them stronger and more self-confident in figuring things out by themselves. Learning to handle and solve complicated issues and add value to the team in their own way is a rewarding experience. At the same time, inner freedom comes with autonomy and responsibility. The more self-confident your team members become, the more responsibility they can han-

https://doi.org/10.1515/9783111325644-008

dle, the more autonomous they become, and the more freedom they will feel and acquire. However, offering your people safety, care, and freedom only is not enough and does not work; in return, you expect responsibility, accountability, and autonomous initiative. To achieve this, the best approach is probably to give trust, manageable freedom, and attainable challenges. Gradually, they can build self-confidence, take on responsibility, and gain freedom for initiative. If successful, your organization may look forward to sustainable growth and impact.

In the next phase, you should share the Playing Field with your team (with the structure, mission, vision, ambition, goals, culture, team members, team play, etc.) to give them a compass for working in your environment. Ask for feedback on how they perceive your Playing Field's implementation, as this stimulates their involvement. Suggest they make their own 6T trajectory plan, incorporating their and your shared interests. Decide on the final 6T plan to integrate it with your Playing Field. This commits. Then support them to perform and create impact with their 6T plans. For this, the 9-elements program may help them.

## 8.2 At the level of a larger organization: the top-down approach

When you are leading a larger unit such as a department, a university, or a research institute, you will only have frequent contact with some of your people. The only way to implement principles of flock leadership is via others. You can start a top-down process and begin by sharing your ideas of governance and leadership with heads of department, group leaders, chair holders, board members, or in a hierarchical structure, the people reporting to you. In our practice with clients, we start with a presentation on flock leadership to acquaint the participants with the advantages of trust, freedom, openness, and responsibility as well as the 3B, 6T, and 9E principles. Further, we use a custom-specific version of our integral governance model (like the comprehensive scheme in Chapter 5, Fig. 5.4) to analyze and share the actual situation in an open atmosphere. The colored-notes approach (Fig. 5.6) brings practical and strategic issues to the table. In this way, one hopes that in addition to some successes and strengths one can be proud of, hidden frustrations and resistances also become apparent. Generally, people gain confidence and become more open when they feel safe to share hidden negative aspects openly in an atmosphere and culture of trust. If successful, it enables a healing process aimed at building confidence in each other. When done correctly, people will feel energetic and proactive after such a session. Having experienced this process constructively, the participants may become enthusiastic of flock leadership and adopt it. They may want to share their experience with their teams – the next layer in the organization – and set up similar sessions in their environment. If successful, implementing the new governance and leadership philosophy trickles down layer by layer in the organization.

The cascading process for the entire department, institute, or company can be facilitated by an external coach or, even better, by internal coaches, such as professionals from the human resources department, and, very desirably, the team leaders reporting to you. When the leader participates, absolute trust is built because people will feel that their leader genuinely supports the process. Otherwise, building trust may be limited to the facilitator and the other participants, who will not know the direct leader's position in the process.

Establishing cross-connections between subgroups in similar stages of translating and implementing the playing field at the level of each team may be helpful. The cross-border element 2.3 of the 9-elements table (Tab. 7.2) is directed at this. One may use the triple set self-leadership assessments to define the required personality roles and styles for each function and match them with self- and 360 assessments. Remember, this is a top-down process. Getting all key people on the same page may prove to be a challenge, but the endeavor is worthwhile if you want to create a trusted and agile organization and are determined to make it work.

## 8.3 Middle-up-down

An alternative and more gradual route to implementing flock leadership principles in your organization is to focus on your emerging key players – the "leaders of the future." This could be in the form of an internal academy, a "Good Sports and Institute Development Program," or a "Master-Your-Future" scheme for the coming generation such as a university's assistant and associate professors. Participants get to know each other in a trusted environment, share experiences, and encourage each other to work along the 3B-6T-9E philosophies. The spirit, atmosphere, and initiatives are shared with the executive level in periodic meetings. As part of the course, the participants create tangible proofs with assignments for them in the organization. There is a good chance that these spread organically throughout your institution. The board may notice that things are changing for the better, get confidence that the philosophy works, and adopt it. From there on, flock leadership may spread more easily through the entire organization or parts thereof.

## 8.4 Bottom-up

Another approach to complement the implementation of policies at a higher level is to make the 3B-6T-9E-philosophy part of your human resources policy or human resource management development plan. One may, for example, start with the newly arriving staff and with promising young professionals already in the organization. This bottom-up approach is directed at encouraging the young staff toward self-leadership becoming – and staying – authentic (3Bs), making their own 6T plan and learning what is needed to achieve personal and organizational goals effectively (9Es). This approach needs support from an experienced person who knows what it takes to become and behave authentically, even when the context is not ready for it, to make people aware of the risks of vulnerability in an arena environment and what it takes to handle this.

# IMPLEMENTING SELF LEADERSHIP

– TOP DOWN: e.g.
  – A RESEARCH GROUP, or
  – AN ENTIRE INSTITUTE OR UNIVERSITY

– MIDDLE UP DOWN: e.g.
  – IN A DEPARTMENT, IN AN ORGANIC WAY

– BOTTOM UP: e.g.
  – COACHING AND ENHANCING
    INDIVIDUALS AT ALL LEVELS

**Fig. 8.1:** Three ways for implementing self-leadership programs in an organization.

## 8.5 Finally

Figure 8.1 summarizes the three approaches for implementing self- and flock-leadership in larger organizations such as a university, department, or a research institute.

In all cases, and if desired, leaders and participants can use the e-learning facility accessible via www.scientificleaders.com, where education programs and assessments on performance roles, intra-/entrepreneurship attitudes, and (self)-leadership styles are available.

The final pages of this book contain scorecards and schemes for use with colleagues, teams, or just by yourself.

# Worksheets

# Score Card
## Your Personal Dimensions

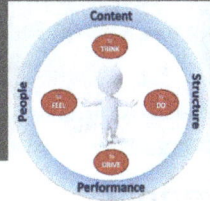

## to Feel

| too little | | | balanced | | | | too much | | |
|---|---|---|---|---|---|---|---|---|---|
| 1 | 2 | 3 | 4 | 5 | 6 | 7 | 8 | 9 | 10 |

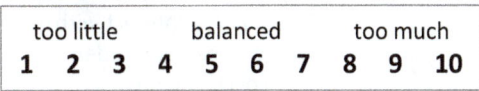

Use of feelings, orientation on people, interaction, networking

## to Think

| too little | | | balanced | | | | too much | | |
|---|---|---|---|---|---|---|---|---|---|
| 1 | 2 | 3 | 4 | 5 | 6 | 7 | 8 | 9 | 10 |

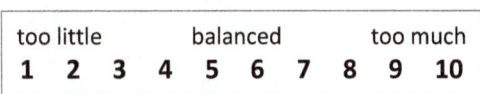

Orientation on content and intrinsic meaning, often in larger context; strategy

## to Do

| too little | | | balanced | | | | too much | | |
|---|---|---|---|---|---|---|---|---|---|
| 1 | 2 | 3 | 4 | 5 | 6 | 7 | 8 | 9 | 10 |

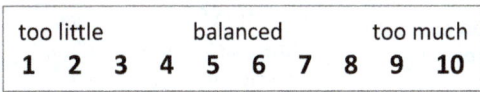

Focus on action and execution of tasks, preferably according to established procedures

## to Drive

| too little | | | balanced | | | | too much | | |
|---|---|---|---|---|---|---|---|---|---|
| 1 | 2 | 3 | 4 | 5 | 6 | 7 | 8 | 9 | 10 |

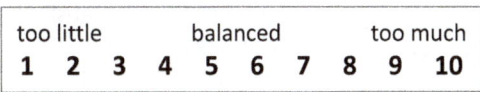

Focus on output, achieving goals, completion of tasks, tangible results within deadline

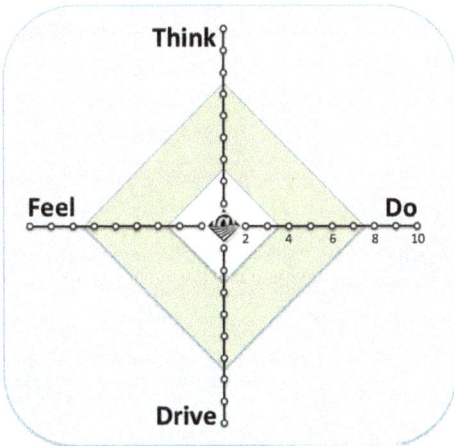

https://doi.org/10.1515/9783111325644-009

# Score Card
## Your Accomplishment Dimensions

SynCat Ac@demy

## People

| too little | | | balanced | | | too much | | |
|---|---|---|---|---|---|---|---|---|
| 1 | 2 | 3 | 4 | 5 | 6 | 7 | 8 | 9 | 10 |

The way you interact with people, supervise students, coach coworkers

## Content

| too little | | | balanced | | | too much | | |
|---|---|---|---|---|---|---|---|---|
| 1 | 2 | 3 | 4 | 5 | 6 | 7 | 8 | 9 | 10 |

Your attention for content and intrinsic value of activities, background knowledge

## Structure

| too little | | | balanced | | | too much | | |
|---|---|---|---|---|---|---|---|---|
| 1 | 2 | 3 | 4 | 5 | 6 | 7 | 8 | 9 | 10 |

Attention for procedures, routines, protocols, deadlines, discipline in your work

## Performance

| too little | | | balanced | | | too much | | |
|---|---|---|---|---|---|---|---|---|
| 1 | 2 | 3 | 4 | 5 | 6 | 7 | 8 | 9 | 10 |

Extent to which you are satisfied with achieving goals, producing output, having impact

# My Present Activities   **1**

## An Analysis of Successes, Concerns and Failures

SynCat Ac@demy

**Assignment:** Take 15 minutes and write down everything that comes to your mind about your activities, whether these concern daily organization or long-term planning, successes, worries about funding, issues in teaching, supervision of students, state of laboratory and infrastructure, safety, position in the department, publications, conferences, your impact in the field, etc.  Don't worry about order, or the nature of issues – small or big, just write in random order what comes up if you think about your work. Try to list at least 12 points for each category.

| What Goes Well? | What Needs Attention? | What Goes Wrong? |
|---|---|---|
| 1 | 1 | 1 |
| 2 | 2 | 2 |
| 3 | 3 | 3 |
| 4 | 4 | 4 |
| 5 | 5 | 5 |
| 6 | 6 | 6 |
| 7 | 7 | 7 |
| 8 | 8 | 8 |
| 9 | 9 | 9 |
| 10 | 10 | 10 |
| 11 | 11 | 11 |
| 12 | 12 | 12 |
| 13 | 13 | 13 |
| 14 | 14 | 14 |
| 15 | 15 | 15 |
| 16 | 16 | 16 |
| 17 | 17 | 17 |
| 18 | 18 | 18 |

# My Present Activities  2

## An Analysis of Successes, Concerns and Failures

SynCat Ac@demy

**Assignment:** Combine the entries in the form 'My Activities – 1' as green, orange, and red numbers (or short descriptions) in the scheme below under the appropriate headings. Is your attention for the important aspects of your work adequately spread or are there blind spots in your approach to management and leadership?

## My Organization:   Why – How – What

*philosophy*

*practice*

*recognition*

| *Vision* | *Strategy* | *Publicity Plan* |
| *Culture* | *Management* | *PR* |
| *Projects* | *Personnel* | *Output* |

*internal*                    *external*

## Resources:

| People | Infra structure |

| Funding |

## Conclusions:

........................................................          ........................................................
........................................................          ........................................................
........................................................          ........................................................
........................................................          ........................................................
........................................................          ........................................................
........................................................          ........................................................
........................................................          ........................................................

# Triggers, Talents, Thrives, Thrills
## on the trajectory to leadership

SynCat Ac@demy

**HOW:**
**6 Trajectory**
to authentic leadership,
high performance
and strong impact

Success

*Your Current Performance*

5.1 **T**rail 1: what can you achieve by yourself

5.2 **T**rail 2: what can be achieved with
support of a good team

5.3 **T**rail 3: how can you create
impact by synergy with a
self-propelling and
empowered team

*Reflection*
*'Deep Dive'*

1. What **T**riggers you?
   Become aware of
2. your **T**alents,
3. your **T**hrives

4. **T**hrill, inspired action:
   personal entrepreneurship
   matched with personal enterprise

6. **T**rack
   monitor and assess each
   trail on accomplishments:
   · Objectives
   · Activities
   · Results

## Thrives
**What are you passionate about?**

..............................................................
..............................................................
..............................................................
..............................................................
..............................................................

## Triggers
**What triggers you in your work?**

..............................................................
..............................................................
..............................................................
..............................................................
..............................................................

## Talents
**What are your most important talents**

..............................................................
..............................................................
..............................................................
..............................................................
..............................................................
..............................................................

## Trails
**What would be important steps towards success?**

*1. On your own* .............................................
..............................................................
..............................................................
..............................................................

*2. With the help of a team* ...............................
..............................................................
..............................................................
..............................................................
..............................................................

*3. With a self-propelling team ?* ..........................
..............................................................
..............................................................
..............................................................

# Team Score Card
## Accomplishment Dimensions

SynCat Ac@demy

Score the three most important members of your team, and yourself.
Is the orientation of the team on different aspects of the work balanced?

### 1.

**People**

| too little | | | balanced | | | too much | | | |
|---|---|---|---|---|---|---|---|---|---|
| 1 | 2 | 3 | 4 | 5 | 6 | 7 | 8 | 9 | 10 |

**Content**

| too little | | | balanced | | | too much | | | |
|---|---|---|---|---|---|---|---|---|---|
| 1 | 2 | 3 | 4 | 5 | 6 | 7 | 8 | 9 | 10 |

**Structure**

| too little | | | balanced | | | too much | | | |
|---|---|---|---|---|---|---|---|---|---|
| 1 | 2 | 3 | 4 | 5 | 6 | 7 | 8 | 9 | 10 |

**Performance**

| too little | | | balanced | | | too much | | | |
|---|---|---|---|---|---|---|---|---|---|
| 1 | 2 | 3 | 4 | 5 | 6 | 7 | 8 | 9 | 10 |

### 2.

**People**

| too little | | | balanced | | | too much | | | |
|---|---|---|---|---|---|---|---|---|---|
| 1 | 2 | 3 | 4 | 5 | 6 | 7 | 8 | 9 | 10 |

**Content**

| too little | | | balanced | | | too much | | | |
|---|---|---|---|---|---|---|---|---|---|
| 1 | 2 | 3 | 4 | 5 | 6 | 7 | 8 | 9 | 10 |

**Structure**

| too little | | | balanced | | | too much | | | |
|---|---|---|---|---|---|---|---|---|---|
| 1 | 2 | 3 | 4 | 5 | 6 | 7 | 8 | 9 | 10 |

**Performance**

| too little | | | balanced | | | too much | | | |
|---|---|---|---|---|---|---|---|---|---|
| 1 | 2 | 3 | 4 | 5 | 6 | 7 | 8 | 9 | 10 |

### 3.

**People**

| too little | | | balanced | | | too much | | | |
|---|---|---|---|---|---|---|---|---|---|
| 1 | 2 | 3 | 4 | 5 | 6 | 7 | 8 | 9 | 10 |

**Content**

| too little | | | balanced | | | too much | | | |
|---|---|---|---|---|---|---|---|---|---|
| 1 | 2 | 3 | 4 | 5 | 6 | 7 | 8 | 9 | 10 |

**Structure**

| too little | | | balanced | | | too much | | | |
|---|---|---|---|---|---|---|---|---|---|
| 1 | 2 | 3 | 4 | 5 | 6 | 7 | 8 | 9 | 10 |

**Performance**

| too little | | | balanced | | | too much | | | |
|---|---|---|---|---|---|---|---|---|---|
| 1 | 2 | 3 | 4 | 5 | 6 | 7 | 8 | 9 | 10 |

### yourself

**People**

| too little | | | balanced | | | too much | | | |
|---|---|---|---|---|---|---|---|---|---|
| 1 | 2 | 3 | 4 | 5 | 6 | 7 | 8 | 9 | 10 |

**Content**

| too little | | | balanced | | | too much | | | |
|---|---|---|---|---|---|---|---|---|---|
| 1 | 2 | 3 | 4 | 5 | 6 | 7 | 8 | 9 | 10 |

**Structure**

| too little | | | balanced | | | too much | | | |
|---|---|---|---|---|---|---|---|---|---|
| 1 | 2 | 3 | 4 | 5 | 6 | 7 | 8 | 9 | 10 |

**Performance**

| too little | | | balanced | | | too much | | | |
|---|---|---|---|---|---|---|---|---|---|
| 1 | 2 | 3 | 4 | 5 | 6 | 7 | 8 | 9 | 10 |

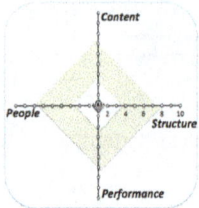

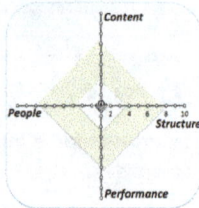

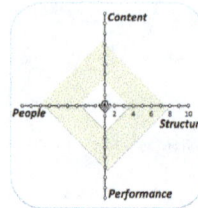

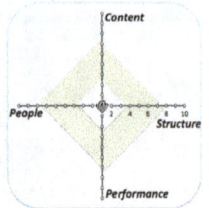

## The Team *(all profiles combined)*:

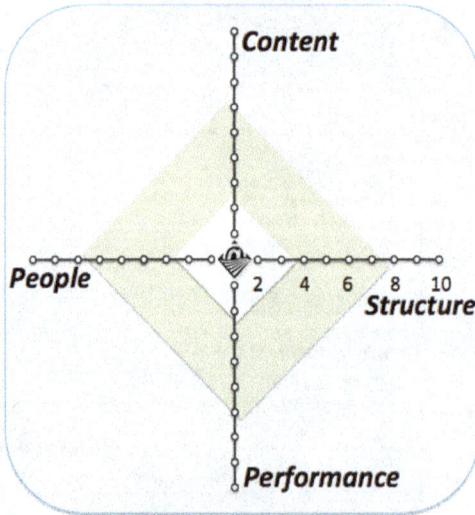

# Team Score Card
## Personal Dimensions

SynCat Ac@demy

Score the three most important members of your team, and yourself.
Is the team balanced?

## 1.

**to Feel**

| too little | balanced | too much |
|---|---|---|
| 1  2  3 | 4  5  6  7 | 8  9  10 |

**to Think**

| too little | balanced | too much |
|---|---|---|
| 1  2  3 | 4  5  6  7 | 8  9  10 |

**to Do**

| too little | balanced | too much |
|---|---|---|
| 1  2  3 | 4  5  6  7 | 8  9  10 |

**to Drive**

| too little | balanced | too much |
|---|---|---|
| 1  2  3 | 4  5  6  7 | 8  9  10 |

## 2.

**to Feel**

| too little | balanced | too much |
|---|---|---|
| 1  2  3 | 4  5  6  7 | 8  9  10 |

**to Think**

| too little | balanced | too much |
|---|---|---|
| 1  2  3 | 4  5  6  7 | 8  9  10 |

**to Do**

| too little | balanced | too much |
|---|---|---|
| 1  2  3 | 4  5  6  7 | 8  9  10 |

**to Drive**

| too little | balanced | too much |
|---|---|---|
| 1  2  3 | 4  5  6  7 | 8  9  10 |

## 3.

**to Feel**

| too little | balanced | too much |
|---|---|---|
| 1  2  3 | 4  5  6  7 | 8  9  10 |

**to Think**

| too little | balanced | too much |
|---|---|---|
| 1  2  3 | 4  5  6  7 | 8  9  10 |

**to Do**

| too little | balanced | too much |
|---|---|---|
| 1  2  3 | 4  5  6  7 | 8  9  10 |

**to Drive**

| too little | balanced | too much |
|---|---|---|
| 1  2  3 | 4  5  6  7 | 8  9  10 |

## yourself

**to Feel**

| too little | balanced | too much |
|---|---|---|
| 1  2  3 | 4  5  6  7 | 8  9  10 |

**to Think**

| too little | balanced | too much |
|---|---|---|
| 1  2  3 | 4  5  6  7 | 8  9  10 |

**to Do**

| too little | balanced | too much |
|---|---|---|
| 1  2  3 | 4  5  6  7 | 8  9  10 |

**to Drive**

| too little | balanced | too much |
|---|---|---|
| 1  2  3 | 4  5  6  7 | 8  9  10 |

## **The Team** *(all profiles combined)*:

## Towards a Comprehensive Plan - 1

SynCat Ac@demy

### Vision:
*Your view on what is needed in larger context*

.................................................

.................................................

.................................................

.................................................

.................................................

.................................................

.................................................

### Mission:
*What you want to achieve in the long term*

.................................................

.................................................

.................................................

.................................................

.................................................

.................................................

.................................................

### Goals:

**Short term**

.................................................

.................................................

.................................................

.................................................

.................................................

**Medium term**

.................................................

.................................................

.................................................

.................................................

.................................................

**Long term**

.................................................

.................................................

.................................................

.................................................

.................................................

## Towards a Comprehensive Plan - 2

SynCat Ac@demy

## *SWOT Analysis*

|  |  |
|---|---|
| ............................................ | ............................................ |
| ............................................ | ............................................ |
| ............................................ | ............................................ |
| ............................................ | ............................................ |
| ............................ *Strength* | *Weakness* ............................ |
| *Opportunity* | *Threat* |
| ...................... | ............................ |
| ............................ | ............................ |
| ............................ | ............................ |
| ............................ | ............................ |

### Conclusions:

_____

_____

_____

_____

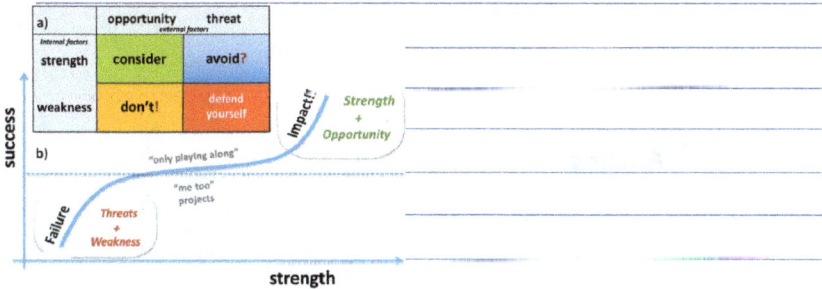

# Towards a Comprehensive Plan - 3

SynCat Ac@demy

## My Organization:  Why – How – What

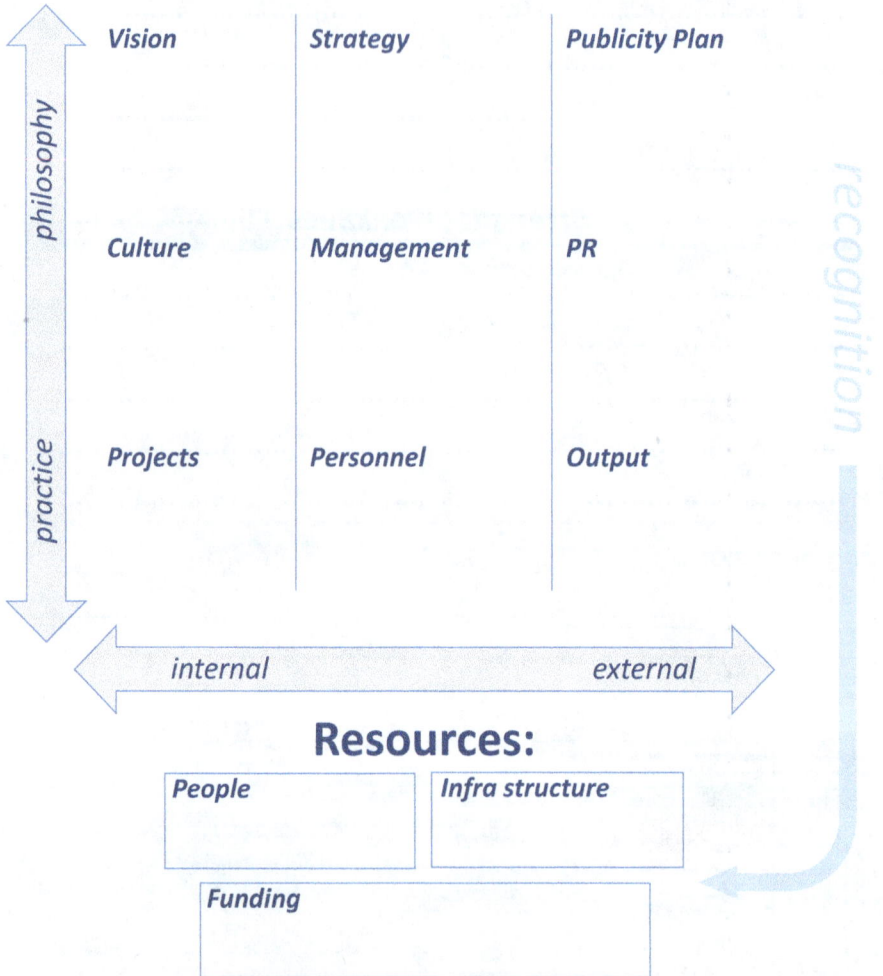

*philosophy*

*practice*

*recognition*

| *Vision* | *Strategy* | *Publicity Plan* |
|---|---|---|
| *Culture* | *Management* | *PR* |
| *Projects* | *Personnel* | *Output* |

*internal*  ←→  *external*

## Resources:

| People | Infra structure |
|---|---|
| | |

| Funding |
|---|
| |

# Index

https://doi.org/10.1515/9783111325644-010

www.ingramcontent.com/pod-product-compliance
Lightning Source LLC
Chambersburg PA
CBHW081526220326
41598CB00036B/6350